DWELLING IN THE AGE OF
CLIMATE CHANGE

For all sentient beings

'Suffering is at once what disturbs order and this disturbance itself.'
Emmanuel Levinas

DWELLING IN THE AGE OF CLIMATE CHANGE
The Ethics of Adaptation

Elaine Kelly

EDINBURGH
University Press

Edinburgh University Press is one of the leading university presses in the UK. We publish academic books and journals in our selected subject areas across the humanities and social sciences, combining cutting-edge scholarship with high editorial and production values to produce academic works of lasting importance. For more information visit our website: edinburghuniversitypress.com

© Elaine Kelly, 2018

Edinburgh University Press Ltd
The Tun – Holyrood Road
12(2f) Jackson's Entry
Edinburgh EH8 8PJ

Typeset in 11/13 Sabon by Servis Filmsetting Ltd, Stockport, Cheshire

A CIP record for this book is available from the British Library

ISBN 978 1 4744 2296 3 (hardback)
ISBN 978 1 4744 2297 0 (webready PDF)
ISBN 978 1 4744 2298 7 (epub)
ISBN 978 1 4744 5217 5 (paperback)

The right of Elaine Kelly to be identified as the author of this work has been asserted in accordance with the Copyright, Designs and Patents Act 1988, and the Copyright and Related Rights Regulations 2003 (SI No. 2498).

CONTENTS

Acknowledgements viii
List of Abbreviations x

Introduction 1
    Exceeding calculation: the ethics of adaptation 1
    Dwelling, hospitality and the Other: foreign questions? 7
    Mess: cultural theory and its 'place' 11
    How to read this book 13

Part A
1 The Adaptation Agenda 17
    Local Adaptation 19
    The climate–development agenda 24
    Adaptation: a politics of place 29
    The ethical primacy of the local 32
    Conclusion: on shaky grounds 40

2 Rethinking Adaptation: An Ethos of Dwelling 42
    Dwelling 44
    Plight 45
    Death 47
    Dwelling-with 48
    Dwelling and the Other 48
    An inviolable home: communal relocation, the Inuit
      Petition and the 'right to a home' 52
    Conclusion: deconstructed grounds 58

3 The Mobility Agenda 61
    Migration as a mega-trend 63

The mobility agenda	66
Common but Differentiated Responsibility and the climate migrant	71
An Other-oriented human right to mobility	74
An Other-oriented protection agenda for climate migrants	77
Mobility as social transformation	82
Conclusion: contested grounds	85

Part B

4 In and Out of Place: The Case of Bangladesh — 89

Welcome	89
Local adaptation: situating climate change	93
A politics of place: land as lifeline	95
The limits of the local	99
Does this require global responsibility? Migration as adaptation	102
In practice: Australia–Bangladesh and the possibility of welcome	110
Conclusion: common grounds	116

5 In and Out of Place: The Case of the Torres Strait Islands, Australia — 118

Haunted house	118
Climate impacts in the Torres Strait Islands	121
Local adaptation, local resistance: roots and rights	123
Migration as adaptation	127
Beyond adaptation: relocation in the context of neo-colonialism	129
Hauntology: political autonomy and the future	136
Conclusion: borrowed grounds	141

6 Hitting the Global North: From Crisis to Recovery to Rebuilding? — 143

Hard rain	143
The 'new normal'	145
Undoing universal vulnerability: domicide and social marginalisation	152
Brisbane, Queensland, 2011: 'made of tough stuff'	153
Katrina, New Orleans, 2005: 'we are not refugees'	158
Resisting domicide	163

Contents

Conclusion: groundlessness	164
Conclusion: A Future, Otherwise Than This	165
Notes	171
Bibliography	181
Index	206

ACKNOWLEDGEMENTS

Many thanks to the University of Technology, Sydney, Australia, for the opportunity to undertake a Chancellor's Postdoctoral Fellow between 2011 and 2015. It was out of this work contract that my book project was able to take root and grow. A number of colleagues at UTS made my time there enjoyable and provided energy and conversation that assisted me with this project. Special thanks to Associate Professor Katrina Schlunke, Dr Virginia Watson and Dr Catherine Robinson. I would also like to thank the postgraduate cohort in Cultural Studies at UTS. The UTS Research Office was both patient and generous, extending my funding for fieldwork, for which I am extremely grateful.

My sincere gratitude goes to Edinburgh University Press, in particular Jenny Daly who was both patient and enthusiastic about my project during the reviewing and revision stages. My sincerest thanks go to the four initial reviewers of the proposal for their considerate feedback. I would like especially to thank the anonymous reviewer who gave their time to provide detailed and generous feedback on the complete monograph draft. This feedback was invaluable.

Professor Nick Mansfield has been a mentor, colleague and friend for many years and I thank him deeply for all the support and philosophical debates we have had. Sincere thanks also to Michael McElhone for very generously giving his time to read drafts of this book; Michael pushed me to write as clearly as I could! Andonea Dickson provided enthusiastic and insightful feedback on an early draft, and Kirstin Steward, Ravi Glasser-Vora, Dr Jess Gifkins, Dr Katherine Carroll, Dr Elaine Laforteza (and Violet), Karen Green and Sophie Martin (and Clemmie) have given immeasurable intellectual and emotional support over the years. Thanks also to Professor Jane McAdam for all of the help with fieldwork she provided, as well as her kind support of my work.

Acknowledgements

Professor Claire Colebrook provided practical support for my work, from the first time we met at a conference, right through to publication.

In 2015, I joined a long-running writing group based in Sydney. I am very grateful to Lena Bruselid, Michael Dudley, Catherine Hickie, Winton Higgins, Lorraine Rose and Maurice Whelan for their critical and thoughtful feedback on this work. They provided a non-expert/non-academic perspective which, I feel, has assisted with the clarity of the final piece.

My thanks to the people in the Torres Strait Islands and Bangladesh, who kindly gave their time to speak with me.

My partner David Tanner made this project possible. When difficulties arose, he was there to help. The support he has provided is beyond the reach of words to convey. His sharp eye was very useful in the editing process too. My sister, Suzi, has always been there to keep me grounded and practical, and my nanna and in-laws have offered the necessary space away from writing. Thank you Gus: I thought of you often when writing this book, a book about the future of our world, a world that you will grow up in.

There are others, many others, who have helped me in different ways. All of my friends – you know who you are. Thank you.

ABBREVIATIONS

CBDR	common but differentiated responsibilities
COP	Conference of Parties
DFAT	Department of Foreign Affairs and Trade
GCF	Green Climate Fund
HLP	housing, land and property (rights)
IOM	International Organisation for Migration
IPCC	Intergovernmental Panel on Climate Change
KANI	Kiribati Australia Nursing Initiative
LDC	least developed country
NAP	national adaptation plans
NGO	non-governmental organisation
TSI	Torres Strait Islands, Torres Strait Islanders
TSRA	Torres Strait Regional Authority
UDHR	Universal Declaration of Human Rights
UNFCCC	United Nations Framework Convention on Climate Change
UNHCR	United Nations High Commissioner for Refugees
WWII	World War II

INTRODUCTION

How to distinguish between two disadjustments, between the disjuncture of the unjust and the one that opens up to the infinite asymmetrical relation to the other, that is to say, the place for justice. Not for calculable and distributive justice. Not for law, for the calculation of restitution ... Not for calculable equality, therefore, not for the symmetrizing and synchronic accountability or imputability of subjects or objects, not for a *rendering justice* that would be limited to sanctioning, to restituting, and to *doing right*, but for justice as incalculability, of the gift and singularity of the an-economic ex-position to others.

Jacques Derrida[1]

EXCEEDING CALCULATION: THE ETHICS OF ADAPTATION

It is tempting to argue that the international response to climate change is devoid of ethics. In one fell swoop I could contend that we have made no progress, are doing nothing and are completely at the whim of an out-of-control global economic agenda that threatens to annihilate us all in the pursuit of a minority of (extremely wealthy) interests. After all, the actions that have been taken globally in the name of mitigation have fallen well short of what is needed to prevent dangerous climate change. Indeed, global emissions of CO_2 have climbed by 2.5 per cent annually since the turn of the new millennium (Friedlingstein et al. 2014). It seems that we are blindly hurtling towards our extinction. It would not be unreasonable to wonder, then, if we have lost our ethical bearings. Naomi Klein has forcefully highlighted our collective inertia, reminding us that climate change is 'an emergency, a present emergency, not a future one, but we aren't acting like it' (2016: 14). Klein is certainly right: the consequences of global warming are real and urgent and we, or at least we in the Global North, continue to be disconnected from this fact.[2]

Insufficient international action does not, however, map onto a lack of available ethical norms. Despite appearances, our ethical imagination is alive. For decades, political leaders, international and domestic NGOs, scientists, activists, religious leaders and others have been developing and implementing extensive frameworks and programmes aimed at addressing the issue, cognisant of the importance of ethics alongside political, economic and legal concerns. The *United Nations Framework Convention on Climate Change* (UNFCCC) 1992 embeds, at the outset, two normative ethical foundations according to which action should be taken: intergenerational justice and distributed responsibility. At the time of its emergence, the UNFCCC was concerned primarily with mitigation, and indeed, it was the springboard for later international negotiations such as the *Kyoto Protocol* 1997, out of which emissions reduction targets were developed. Article 3(1) states unequivocally that:

> The Parties should protect the climate system for the benefit of present and future generations of humankind, on the basis of equity and in accordance with their common but differentiated responsibilities and respective capabilities. Accordingly, the developed country Parties should take the lead in combating climate change and the adverse effects thereof. (UNFCCC 1992)[3]

Within the UNFCCC, 'common but differentiated responsibilities' (CBDR) has been shorthand for climate ethics and justice since its emergence. These ideals are there to guide decision-making and action. Such principles speak to the duties we have towards one another and acknowledge, in broad brush strokes, the inequities that characterise the issue of climate change (identified in the shorthand of North–South negotiating positions). At present, the principle of responsibility for historical emissions is at the forefront of ethical theory related to climate change (Gardiner et al. 2010). In other words, nation-states have different degrees of responsibility for reducing carbon emissions reflected in distinct national targets. By assigning obligations to governments based on their historic greenhouse gas emissions, the above-mentioned ethical norms can, in principle, provide a way to hold states accountable.

Introduction 3

What role these, or other, principles may play in relation to climate *adaptation* policy and action is still a work-in-progress. Despite the UNFCCC acknowledging inevitable climate impacts, climate change discourse has unhelpfully placed mitigation and adaptation in a binary: one is either preventing or responding (Head 2010: 236). This logic has meant that adaptation came late to the negotiating table, only making a meaningful appearance from the *Kyoto Protocol* 1997 negotiations onwards. When adaptation did begin to gain traction, assigning responsibility for contemporary and projected adverse effects became an urgent but complex topic (Dellink et al. 2009; see also Müller et al. 2009).

The usefulness of CBDR has met some insurmountable obstacles in relation to adaptation. Why? Chukwumerije Okereke explains that developed countries 'insist that the ground for compensatory justice is extremely weak given the difficulty, perhaps the improbability, of establishing clear lines between cause and effect' (2010: 465). Causality – that is, this action led to that effect – continues to be a contentious point because it requires some admission of responsibility. In law, there are three steps involved in establishing obligation: the designation of a duty, a breach of the duty, and from there evidence of causal impact. An acknowledgement that some duty has been breached comes with calls for financial compensation and legal culpability. The financial costs of adaptation are estimated to be between US$70 and US$100 billion annually from 2020 (World Bank 2011). As a result, few industrialised countries are keen to accept the costs of taking responsibility.

Consequently, the extension of CBDR principles to climate adaptation has largely failed to come to pass in both principle and reality. While major adaptation financial schemes such as the Green Climate Fund (GCF) were designed in order to enable 'developed to developing' country financial assistance, the voluntary nature of the scheme combined with 'what appears to be a retreat from the CBDR framework' (Vanderheiden 2015: 37) demonstrates a wide gulf between ethics and politics in global negotiations. At best, applications of CBDR have been haphazard: while the Least Developed Countries (LDC) Fund has had some success, the criteria for funding in relation to the Adaptation Fund have been subject to contestation and therefore it has been much slower to operationalise (Honkonen 2009: 260).

While mitigation is focused on carbon emissions, adaptation is a broad platform for action and consequently the route to accomplishing it is more complex still. At first glance, adaptation may refer simply to the ability to alter or change in order to fit with one's environment; an understanding that reflects its moorings in evolutionary thinking. As I demonstrate in Chapter 1, the baggage of this linguistic debt still remains, but has been transformed to take account of the social factors that inhibit or enhance one's capacity to adapt. What remains a strong feature of our adaptation discourse is a focus on local habitation. As a result of this, adaptation policies have been filtered through development agendas and focus on maintaining relationships to local place through agricultural, infrastructural and social development support.

Adaptation is not merely a descriptor. Rather, it is a thoroughly prescriptive term. Particular human practices are given the label of 'adaptive' while others remain fiercely contested. Primary amongst the contentious practices is the ancient act of migration or, more broadly, human mobility (encompassing displacement, migration and relocation). There is still debate as to whether or not mobility should be mainstreamed into adaptation platforms, with a formal consensus remaining out of reach for the time being. Nonetheless, forms of climate-induced mobility are already occurring across the world. In the Pacific, the Carteret Islander community has been in the process of communal relocation for a number of years, while Indigenous Alaskan communities are also undertaking the slow and difficult work of setting up communities in new places. In Bangladesh, the case study explored in Chapter 4, extensive rural to urban migration has been correlated with environmental impacts such as flooding, drought and cyclonic activity.

History gives us a picture of a complex and global migration system; one overwhelmingly skewed towards the interests of the privileged. Moreover, history may provide us with many instances of displacement and forced migration, but the scale of climate-related disturbances threatens to exceed these precedents. The International Organisation for Migration (IOM) estimates that approximately 200 million people will be displaced as a result of climate change by 2050. We can see, then, that the risk of focusing predominately on the 'local' in our adaptation platforms is that forms of mobility, whether 'forced' or 'voluntary' in nature, are ignored or relegated to the fringes of international decision-making.

This is precisely what has unfolded. Globally, an enormous and increasingly problematic gap in norms is evident, raising serious concerns regarding questions of substantive justice, that is, of rights and responsibilities. At the time of writing in 2017, there were no binding international frameworks for climate mobility, whether categorised as 'forced' or 'voluntary'. Pre-existing protection instruments do not accommodate climate migration, while general human rights associated with mobility are rather scant (as will be discussed in Chapter 3). While internal displacement can be accommodated under the current *IASC Operational Guidelines on the Protection of Persons in Situations of Natural Disaster* 2011, many countries are not signatories. Relocation dialogues and activities are taking place in a piecemeal fashion and often inadequately funded (Semuels 2015). Protection for those who are forcibly displaced across borders relies upon pre-existing human rights paradigms which, while insistent that individual dignity is paramount, are unable to secure refuge for those who have lost their homes to disasters.

Are these shortcomings reflective of a political problem or one more serious still? That is, is this situation merely one of political significance or does it reflect the weaknesses of our ethical imagination too? Are our ethical frameworks up to the task? For some climate ethicists the paradigms of distributive and intergenerational justice provide a sufficient response to the conceptual knot 'causality' provokes, and as such are suitable reference points for our adaptation planning. However, as I will argue in this book, with regards to complex issues of adaptation and human mobility, CBDR may not be able to produce the sorts of social and political transformations we need. This is because, primarily, such norms are prescriptive, universal and top-down. While this may sound unproblematic in the sense that it promises the application of some form of 'justice' and/or accountability through the nation-state, many thinkers have pointed to the problematic dualisms between citizen and foreigner, and self and Other, that normative theory perpetuates (Jaggar 2013: 441). It perpetuates dualities because 'politics' comes from the Greek *politicos*, meaning 'of, for or relating to citizens'. Human rights doctrines and international law have, of course, come along and challenged the exclusivity of citizenship. Nonetheless, at present human rights require the model of the sovereign state in order to be recognised and applied. Moreover, CBDR relies upon the assumption that responsibility is calculable. This calculative logic is often skewed towards the interests of the powerful rather than the disadvantaged and thus not aimed at a 'just' distribution.

As I will demonstrate in this book, continental theory offers an alternative vocabulary able to help us navigate the mountainous ethical terrain of a climate change era. William Schroeder tells us that continental thinking has a substantially distinct 'tone' (2013: 461). The analytic traditions are able to create change via institutional reforms, but they assume an 'inherent egoism' in the individual and from there they are designed to provide a 'policing function' that keeps this innate egoism of both individuals and institutions in check (2013: 483). Continental thinkers well appreciate that self-interest and egoism reign supreme (especially at present), but see this as a product of historical and social conditions rather than a reflection of the 'truth' of human nature. For continental ethics, 'the goal, in effect, is not to police or circumscribe, but to awaken and enliven – to energise people to ethical creativity' (2013: 483). Awakening needs others: a social body to sustain itself. Consequently, it is only through personal and social transformation that institutions can possibly 'awaken' too. While I agree with the motivations of equity and justice that inform normative or analytic ethical models, this book draws on continental philosophy in order to offer an alternate lens through which to articulate our ethical responsibilities. It is not, however, a matter of replacing one body of knowledge with another. I am not presenting an antagonistic position with regards to normative ethics. Rather, I am suggesting that continental theory has something important, unique and radical to offer us.

Continental thinkers tend to elevate the significance of the social, cultural and affective in the cultivation of ethical and political action. As Alison Jaggar puts it:

> Individuals with relational moral identities are unlikely to make a sharp separation between their own interests and those of others; they are more likely to be moved by considerations of particular attachment than by abstract concern for duty, more by care than by respect and more by responsibility than by 'right'. (Jaggar 2013: 441)

Jaggar's articulation of relationality emphasises the impossibility of separating self from Other, and from this the demand to think first in terms of responsibility rather than right, and incalculability rather than calculation. Continental thinking recognises that the tough ethical work is in negotiating responsibility within the complex historical and cultural era we find ourselves in. From this, we derive a spirit of ethical creativity informed by our unavoidable situatedness. We need ethical thinking that is responsive to the conditions of impermanence, instability and constant movement. As I will demonstrate, deconstruction offers this type of thinking. While analytic thought can work when our grounds are assumed, deconstructive thinking understands that we are fundamentally groundless and thus perpetually responsive to a world in flux. It is the figure of the migrant that brings this productive instability into sharp focus.

DWELLING, HOSPITALITY AND THE OTHER: FOREIGN QUESTIONS?

Given the reticent attitude of the Global North to commit to a framework of CBDR in relation to the current development-focused adaptation agenda, it is difficult to imagine the extension of these principles to solutions related to human mobility. This is especially because the movement of people tends already to be cast as a security concern rather than a moral one (Huysmans 2006; Boas 2015; Vaughan-Williams 2015). Moreover, as has been extensively argued, 'causality' is exceptionally nuanced with reference to migration (Foster 2007). Is it, specifically and primarily, the human-induced changes in the climate that are responsible for the instances of migration? No, it is never solely the environment. Our ethical response to climate change must accommodate this fact and embrace the challenges that arise in a messy human world that cannot be reduced to linear causality or bounded, unrelenting promises of 'security'. Despite the strong religious principles of hospitality and refuge available for use, and perhaps because of these tensions, the organisation of human mobility has been cast as a political and economic consideration rather than an ethical one (see Ahn 2010; Wilson 2013). Consequently, the nation-state retains authority over the conditions of legal belonging. Notable expressions of the rights of the visitor come in Kant's cosmopolitan articulation of hospitality, which argued for the right of the stranger to be treated kindly upon visitation (see Derrida 2000; Brown 2010). Meanwhile, post-World War II (WWII) protection agendas focused on extending refuge have retained their narrow and problematic political persecution criterion.

Since the turn of the new millennium, Jacques Derrida's attention to questions of hospitality, ethics and politics has spurred a substantial and transdisciplinary interest in the language and practice of hospitality for the contemporary world. Deploying the deconstructive logic Derrida offers, scholars in disciplines ranging from cultural studies, philosophy, human geography, international relations, political theory, law, theology and sociology have written extensively on the ethical and political considerations that arise in relation to human migrations (see Pugliese 2002; Baker 2013; Friese 2004; Bulley 2006, 2017; Clark 2011; Fagan 2013; Rosello 2001). A normative way of framing the issue of hospitality might be to ask: 'what can a stranger expect . . . what can he demand and which rights are accorded to him?' (Friese 2004: 67). These are important questions and ones that I engage with over and again throughout this book: they are concerns regarding what we *ought* to do about 'rights' and 'duties', expectations and realities. However, drawing on the resources of the continental tradition, further questions call upon us to respond. Indeed, part of the ethical task we are charged with, when we come from a poststructuralist tradition, is the critique or deconstruction of sovereignty, both with reference to the nation-state as the dominant political community of our times, and in relation to the individual 'self' who is constituted as a 'sovereign subject' (see Campbell 1998).[4]

At first glance, the question of sovereignty may appear to be quite distinct from that of hospitality or, indeed, dwelling, adaptation, or even ethics. Yet these terms are always already interrelated. Philosopher Heidrun Friese articulates the conceptual reach of hospitality, arguing that:

> The various 'languages of hospitality', which arrange different concepts of the relationship with the Other and mark the site of hospitality, involve the questions of territory and border, of private and public spaces, and they entail the question of what is considered common, as well as concepts of belonging, membership, citizenship, and exclusion. (Friese 2004: 74)

Undoing the supremacy of the sovereign self is an ambitious aim. In this book, I work with the thinking of Martin Heidegger, Emmanuel Levinas and Jacques Derrida (amongst others) in order to move towards this project. Suffice to say that at the national level, the challenge to sovereignty is always in progress regardless of what we like to think or do. The inevitability of human movement defies the logic of the orderly or 'legal'.

Looking back in history, we can see that the issue of migration has provoked crises for human societies, with few enduring ethical principles able to resolve the tensions that inevitably emerge. The tensions provoked by the migrant are, to some extent, traceable to the very language of hospitality that comes to frame host–guest relations. Immediately, 'hospitality' gives rise to 'ambivalence' as seemingly stable dichotomies such as 'hospitality and hostility, proximity and distance, belonging and being foreign, inclusion and exclusion' are placed in contention (Friese 2004: 68).

What Derrida refers to as the 'foreigner question' plays a pivotal role in the deconstruction of sovereignty (Dufourmantelle and Derrida 2000: 3). When we turn towards the migrant, we are turning over a question central to humankind, the question of 'where?' (2000: 52–4). Where do I belong? Where can I go? Where will we go then? Where is my place in this world? The question 'where?' is not only for the one on the move or trapped in the transient and precarious cycles of immobility, displacement, migration and dispossession. Rather, it is a question for each of us, especially those of us with a place of safety. Why? It is because the foreigner's cry for help calls upon me to respond: to hear their struggle and their suffering. As Levinas puts it: 'Suffering is at once what disturbs order and this disturbance itself' (1998: 78).

The 'foreigner question', as Derrida puts it, makes demands upon those of us who are able to say, 'I am at home.' This question 'puts me in question', writes Derrida drawing on Levinas (2000: 3). When I am put into question in this way, 'the question of the foreigner ['where?'] as a question of hospitality is articulated with the question of being' (Derrida 2000: 9). When confronted with the foreigner's question 'where?' I am compelled to offer 'here, please come in'. More radically, I am compelled to accept, as Levinas so courageously did, that the stranger 'disturbs the being at home with oneself' (1969: 39).

To be at-home with oneself is to dwell. I have selected this term – dwelling – as the gathering point for a cluster of themes that animate the book. On the one hand, for Heidegger, dwelling is a 'task', an orientation in the world, as well as an impetus to act (to *care*) (1978a; Rose 2012). On the other hand, Levinas asserts that dwelling is actually an act of possession, an attempt at excluding the Other (1969). To be at-home is to refuse ethics in the name of self-contentment. Thus he calls this dwelling into question from the beginning. Our dwelling – our home and our being-at-home – is always already open to contestation by the Other: our self-contentment always troubled from within and without.[5] In my reading, Derrida provides a perspective expansive enough to hold both positions as important. The value in a deconstructive approach to dwelling is the way it takes us to the grounds of our home and reveals to us the primary dance we engage with. This is the movement between dwelling as a claim, dwelling as an act of possession, and dwelling as an opening towards the Other that 'puts in question the world possessed' by 'me' (Levinas 1969: 173).

So far, I have cast hospitality and dwelling in terms of ethics. However, as a set of practices enmeshed in social and historical conditions they are also irrevocably political; imbricated in power (Bulley 2017). What then do ethics and politics have to do with one another? The relationship between ethics and politics is yet another instance of what I earlier referred to as 'ambivalence'. It is tempting to hold the two apart, preserve the purity of ethics while condemning politics to ruin, but this is an impossible position (see Fagan 2013). Already in Levinas's work we see that ethics and politics are entangled with one another. Levinas refers to the priority of ethics in his seminal work *Totality and Infinity*, contrasting it with the violence of politics, only to undo any discrete relation between the two by declaring that the Third party (the fact of more than one Other, more than one stranger), opens the question of justice, which is there from the beginning:

> The third party looks at me in the eyes of the Other – language is justice. It is not that there first would be the face, and then the being it manifests or expresses would concern himself with justice; the epiphany of the face qua face opens humanity. (Levinas 1969: 213)[6]

Put simply, engaging with the foreigner is so primal, so pivotal to the human experience that it is constitutive of being itself. As hospitality, it is the original ethical relation and, in a matter best understood as primordial, it interrupts our dwelling. More than this, as I will argue in Chapter 3, it is a core political issue, requiring that we continually critique our institutions and practices in the name of justice as the Third. Out of an understanding of the primacy of the foreigner's question, we can begin to reorient contemporary 'rights' paradigms so that the

Introduction

notion of responsibility animates their realisation and spurs us to create better, more just systems (Levinas 1998; Derrida 2001b). This is what I refer to as 'Other-oriented ethics of rights' throughout this book.

Despite the primacy Levinas and Derrida assign to the 'foreigner question', dominant political theory has cast the foreigner question 'where?' as secondary to the figure of the citizen; a nuisance or 'unfortunate phenomenon' (Nail 2015: 26). Basic as it may seem, 'where?' is the most important question we must face in relation to mobile human populations in the era of climate change. Where do I belong? Where can I go? Where is my home, my place of dwelling? Where, indeed, is my place in this world? What does it mean to dwell? These questions demand our attention; they present themselves with a force that suggests a 'non-postponable urgency' (Levinas 1969: 212).

MESS: CULTURAL THEORY AND ITS 'PLACE'

As I noted above, I have chosen to use the concept of 'dwelling' as the gathering point for my analysis. Sociologist John Law's provocative and transgressive book *After Method: Mess in Social Science Research* provides an illuminating notion of what is happening when one gathers:

> It connotes the process of bringing together, relating, picking, meeting, building up, or flowing together. It is used to find a way of talking about relations without locating these with respect to the normative logics implied in (in)coherence or (in)consistency. (Law 2004: 160)

Certainly, I have given every effort over to being as clear and cogent as possible with the arguments that populate this book. However, Law's articulation of gathering as analyses that defy a singular location of coherence is pertinent to my undertaking. A neat definition of dwelling will not emerge from this book. What dwelling 'looks like' in various cultural contexts is not the purpose of my discussion.[7] Dwelling, as I understand it, will take on something of a deconstructive dance (explored in Chapter 2). This dance will not reveal to us an end point or set of answers to the crises we are faced with. What it will offer is a conceptual vocabulary, or linchpin, robust enough to compel social and political action aimed at a reorientation of our relationship to the earth, ourselves and Others.

My 'home' discipline, Cultural Studies, is similarly ambivalent to a singular location. Implicitly interdisciplinary, cultural theory has been referred to as an 'outlaw' discipline that borrows extensively from other, more 'traditional' modes of academic inquiry (White and Schwoch 2006: 3). Stuart Hall has written of cultural studies as a 'conjunctural practice' that emerges and morphs according to specific social and political realities (1990: 11). Its objects of study, methods and theoretical debts are exceptionally diverse, with ongoing debate in the community as to whether the discipline has a canon (Grossberg 2006; Nelson and Gaonkar 1996). The transitive and kaleidoscopic nature of cultural studies has implications for thinking through its academic (and general public) audience. Where does this book belong?

This book draws on work in cultural theory, international relations, political theory, philosophy, human geography, ecohumanities, media studies, law, sociology, climate change science, religion and even human evolution. It is informed by poststructuralism, deconstruction, feminism, postcolonialism, ecology and scientific endeavours. Its methods include 'deconstruction' (as uncomfortable as this is in relation to the designation as a method),[8] discourse analysis, philosophical inquiry, and even a dip into ethnography and qualitative methods (interviews). In other words, as an academic enterprise, this book invites its scholarly audience and general readers to negotiate hospitality. As Friese writes, the register of hospitality extends to 'academic practices which are to re-present social worlds. Thus what is at stake is not only the thinking *of* hospitality, but thinking *as* hospitality' (2004: 74).

For Law, the world must be understood to be a 'generative flux that *produces* realities' (2004: 7). Part and parcel of this view is the assumption that the world cannot be tidily arranged into structures (2004: 7). Instead, it is useful to understand it as

> filled with currents, eddies, flows, vortices, unpredictable changes, storms, and with moments of lull and calm. Sometimes in some locations we can indeed make a chart of what is happening around about us. Sometimes our charting helps to produce momentary stability. (Law 2004: 7)

It is within this mess that I undertake my project and attempt to find a point of 'momentary stability', knowing that this will invariably be displaced.

Introduction

HOW TO READ THIS BOOK

There are a number of ways that you can approach this book. In the first instance, you can read it from cover to cover. It is intended to tell a story that connects each of the six chapters. Having said this, it is possible to read each of the chapters as stand-alone pieces. While there are arguments that are woven throughout the chapters and connect each to the other, the sections tend to focus on developing a key idea or case study.

In the first chapter, I examine the political and ethical discourse of local adaptation. Providing an overview of the predominance of the 'local' in our adaptation planning is vital to understanding the way in which human mobility concerns have been positioned in climate politics. Local adaptation is readily accommodated by development agendas, concerned as they are with sustaining the relationship between identity and place.

The second chapter provides the groundwork for an 'ethos of dwelling' that I argue is sorely needed in an era which is witnessing the re-emergence of pervasive protectionist national agendas. This ethos of dwelling is developed out of Heidegger's, Levinas's and Derrida's work. Heidegger's lesser-known piece *Building Dwelling Thinking* provides fertile soil for an elaboration of our responsibility to care for the earth. Bringing Levinas into the conversation allows for an emphasis on our responsibilities to Others with whom I share this dwelling place. In turn, an ethos of dwelling has at its core a concern for the Other's welfare, as well as one's own. The case of the Inuit Petition, as well as the communal relocation of the Newtok peoples in Alaska, is explored in order to demonstrate the need to simultaneously call for the right to a home while continuing to critique expressions of dwelling-as-possession.

In Chapter 3, the emergence of the mobility agenda in adaptation negotiations is described. The promising work of the Advisory Group on Climate Change and Human Mobility in their report *Human Mobility in the Context of Climate Change* (2015) is emphasised. Alongside this, I provide a critical engagement with the centrality of human movement in historical and contemporary culture and politics, as well as with the need to develop a Levinasian Other-oriented ethics of rights which reflects this reality. Taking my inspiration from Thomas Nail's work *The Figure of the Migrant* (2015), I advocate for a shift that elevates the figure of the migrant, the dispossessed and the displaced in our political and ethical thought.

Chapters 4, 5 and 6 focus on how an ethics and politics of adaptation-as-dwelling play out in Bangladesh, the Torres Strait Islands and the Global North, respectively. Chapter 4 focuses on articulating 'rights to a home' in local Bangladeshi adaptation policies. In the second section of the chapter, I shift to the 'responsibilities of hospitality'. Drawing on Derrida's work, I discuss the political implications of hospitality and the need to develop international models to welcome climate migrants or exiles.

When we open ourselves to the primacy of the social and cultural challenges of climate change, we simultaneously open ourselves up to the intangible aspects of belonging, place, identity and loss. Chapter 5 deploys Derrida's work on ghosts and hauntology to discuss the intangible and tangible cultural and social impacts of climate change in the Torres Strait Islands, a string of islands located at the northern-most tip of Queensland, Australia. I argue, with others, that hauntology is Derrida's reframing of intergenerational justice, presented in a philosophical language which understands the 'ghost' as a force confronting us with profound responsibilities to the past and the future.

The final chapter shifts the emphasis from the globally marginalised and instead analyses the discourses of universal vulnerability proliferating in Global North states, such as the US and Australia. Taking the events of Hurricane Katrina in New Orleans, and the Queensland floods in Australia, I examine the politics of dwelling and displacement as they play out in 'climate crises' in the Global North. In the first instance, I contend that we must critique the dominant narratives of trauma, loss and resilience at play in the Global North. This is because there is a tendency to 'universalise' the impacts of climate change. Secondly, an 'ethos of dwelling' in the Global North must be more attuned to the inequalities of power existing socially and politically, as well as the profound impacts this has on issues of displacement, migration and internal relocation. To close, I suggest that an Other-oriented ethics must guide disaster preparedness, response and recovery policies.

PART A

Chapter 1

THE ADAPTATION AGENDA

'We lost our home, which means the familiarity of daily life.'
Hannah Arendt[1]

We are a world on the move, with large numbers of people migrating for work or the promise of a better future. Indeed, 2015 marked the highest recorded number of international migrants, with 244 million people residing in a place other than their country of birth (IOM 2015a). However, not all of this movement is voluntary and the causes and conditions that give rise to forced mobility are proliferating. In 2015, forced displacement reached its highest levels since the end of WWII, with over 65 million people left without a secure home. Just over 20 million of these people met the criteria for refugee status. An additional 19 million people were displaced by natural disasters across 113 countries (2015a: 8). Meanwhile, 'irregular' migration is on the rise, and across the world asylum claims are coming in daily. To add further complications to this, the distinction between forced and voluntary is frequently difficult to maintain (Piguet et al. 2011).

There is agreement that climate change will exacerbate existing vulnerabilities and likely push more people to move, either internally or across borders (Nansen Initiative 2015). Yet there is no consensus on how to respond to such an issue. Indeed, the nexus of climate migration brings together two of the more explosive contemporary political topics: human-induced climate change and human mobility. The highly politicised and securitised nature of migration matters means that it has tended to be marginalised in climate politics. Certainly, climate adaptation dialogues have only just started to formally acknowledge the significance of human mobility both as a potential problem and as a solution or adaptive response. Predominately, however, adaptation refers to practices and policies aimed at sustaining a local relationship between people and place. Indeed, when viewed this way, adaptation is fundamentally a political construction.

What does it mean to say that climate adaptation is a political issue? Instead of trying to determine 'what' politics is, our attention falls instead on the construction of something as political; the 'how' of politics. How does something come to be political? Or, as cultural theorist Jodi Dean puts it, 'What does it mean for something to be political?' (2000: 1). Adaptation is political in multiple ways. In a traditional analysis of what politics is, we can see that climate adaptation is political at the level of governance and decision-making, in relation to how power and resources are distributed and in the manner that responsibility is assigned. In addition to these normative understandings, we can view adaptation as a political project by examining how the meaning of adaptation itself is bound up with the maintenance of unequal relations of power. We can highlight how power comes to shape identities, places, representations and interventions. Within this alternative articulation of the 'political', everything from the language used to convey meaning, to the governance of social bodies, is infused with power and as such it is political.

In this chapter, I argue that adaptation is a political discourse concerned with reaffirming specific understandings of the relationship between place and identity. Derrida refers to the binding of people and place as an 'ontopology'. By this he is referring to the 'ontological value of present-being to its *situation*, to the stable and presentable determination of a locality, the *topos* of territory, native soil, city, body in general' (Derrida 2006: 103; see also Bulley 2006). In other words, there is an assumption that we can stabilise our understanding of the self or sovereign by assigning it an ontological value of presence in a particular place. Climate adaptation is concerned, above all else, with keeping people in their original homelands and as such is an ontopology. In so doing, there is a desire to order and stabilise the world; to remove or deny the incoherency and disorder that marks any ontological claim (Derrida 2006). In this instance it is the inherent possibility of placelessness that makes impossible the self-presence longed for, a point I will explore in due course. For now, shaped by this arbitrary correlate between place and self, the practices that come to constitute an adaptive response (and are therefore fostered and economically supported) are subject to political decisions and have real-world implications for how we view climate-induced mobility. Is migration a form of adaptation or a failure to adapt?

However, emphasising the importance of the 'local' in adaptation planning is also part of a significant ethical discourse which seeks to empower disadvantaged communities throughout the world. Before I examine the specifics of climate migration in Chapter 3, it is vital to appreciate the complexity of the political discourse of local adaptation. It is not simply about keeping people contained, but also about enabling sustainable relationships between people and place. Thus, this chapter attempts to hold together an appreciation of the vital role of local development agendas in climate adaptation, as well as offering a critical perspective of the limits of such an approach. I suggest that the dominant focus on the 'local' is misguided given the reality of large-scale displacement and migration already under way.

LOCAL ADAPTATION

Adaptation has entered common parlance as a term that connotes the ability of something or someone to fit into or adjust to a specific context. Yet its linguistic history reveals some difficult conceptual knots that continue to haunt its application. Already by the early twentieth century the notion of adaptation was contentious. The trouble was that it tended to posit inherent capacities towards survival or extinction. The French biologist Lucien Cuénot argued that adaptation was fundamentally 'a "frightening question" because of the philosophical and metaphysical considerations that it presupposes' (Simonet 2010). This contention arose because Darwin promoted the idea that adaptation was part of the law of nature. As part and parcel of natural law, adaptation was conceived of as both out of the control of human will, as well as incompatible with the belief in a divine power. In other words, adaptation was regarded as random.

This evolutionary legacy brings with it an understanding of 'change' as, at least to some degree, predetermined or biologically coded. Organisms find themselves thrown into an environment in which they are either genetically well equipped to survive or else at great risk of extinction. The continuation of the species relies on the painstakingly slow biological adaptations that may or may not occur over millennia, not the choices that the living beings make during their limited lifespan, or the grand plan of a deity pulling the strings from on high. At its worst, an evolutionary take on adaptation was adopted by Social Darwinists and used to justify forms of colonial exploitation and violence against Indigenous peoples throughout the world who were said to be genetically inferior or 'maladaptive' and therefore doomed to extinction (Head 2010: 234).

In evolutionary terms, the drive for survival may be able to explain why certain adaptations occur. However, it cannot give us a rich understanding of the lively and relational world that we actively participate in and alter as human beings. Thus the explanatory power of evolutionary biology is not designed to provide a platform for approaching the social challenges of anthropogenic climate change. Climate change may have initially been conceived of as a scientific problem but it has become an issue that challenges us to re-examine our ethical and political paradigms. Human-induced climate change requires a more dynamic understanding of adaptation as a process involving multiple overlapping biological, environmental, social, political and psychological factors. How we understand and apply the concept of adaptation reveals to us what we value and how we conceive of our relationship to others and the earth. Thus, a primary question emerges: whose adaptations are we supporting and how?

As we have moved away from a purely evolutionary account of adaptation, its meaning has been revised and expanded. Adaptive strategies are now viewed as flexible, rather than fixed, as we learn to come to grips with the contingent nature of all things (Head 2010: 235). That is, the future is not predetermined by our genetic inheritance. Moreover, myriad conditions give rise to complex formations which inherently change over time. Within the humanistic discipline of sociology, for instance, adaptation refers to the 'reciprocal actions' of the environmental and social domains (Simonet 2010). From a psychological perspective we are attuned to the 'unceasing interaction between Man [sic] and the ever-changing world within which he evolves' (Simonet 2010).[2]

An interactive approach to adaptation, which does not make use of evolutionary logic, is evident in the UNFCCC's working definition:

> adjustments in ecological, social or economic systems in response to actual or expected climatic stimuli and their effects or impacts. It refers to changes in processes, practices, and structures to moderate potential dangers or to benefit from opportunities associated with climate change. (UNFCCC n.d.)[3]

What is striking about this definition is its inclusion of human agency. The UNFCCC promotes a humanistic understanding of adaptation which centres human responsibility and capacity. The message? We are not simply at the mercy of evolution! Successful adaptation requires co-operation. As a result, adjustments can consciously be made in advance of the arrival of dangerous climate change.

Already, adaptation is political. It is a social issue, not a strictly biological one. It is humans who alter ecological systems, as well as economic and social realms. These modifications occur at the level of practices, institutions, processes and structures, and can be defensive or opportunistic. Either way, adaptation is still all about change aimed at fitting in with one's environment. These alterations are always bound up with political, social, cultural and economic processes. Underpinning our adaptive choices are ethical principles concerning our relationship to the earth and to others. Co-operative human action and considered decision-making are as significant as evolutionary factors in determining how adaptation will unfold. Updated conceptualisations of adaptation by the Intergovernmental Panel on Climate Change (IPCC) make this much clearer:

> In human systems, adaptation is the process of adjustment to the actual or expected climate and its effects which seeks to moderate harm or exploit beneficial opportunities. In natural systems, adaptation is the process of adjustment to the actual climate and its effects; human intervention may facilitate adjustment to the expected climate. (IPCC 2012: 3)

So, humans can actively adapt, and find themselves caught up in much larger processes of ecological change that may produce genetic or biological alterations.

We may be able to come to a consensus about the bio-social factors that constitute the meaning of adaptation today. However, not every action is classified as adaptive, even if it leads to personal or social improvement. One key focus of this book is how the processes of recognising a human practice as adaptive (or not) play out with reference to diverse forms of mobility. How does the contemporary discourse of adaptation deal with issues such as chronic displacement from one's home? What about the need, or even wish, to migrate in order to find a home able to sustain one's life? Finally, how does adaptation discourse begin to offer a language for engaging with the loss of culture, land and identity that may accompany, for many, migration and communal relocation?

Language is an expression of power, with meaning caught up in the political and social relations that are operating in any given context. As Derrida (1981) points out, meaning is arbitrary and contingent; it is not inherent and fixed. We can see this above with regards to the linguistic evolution of adaptation. We may be able to stabilise meaning through consensus and refer to 'adaptation' with some shared understanding, but this is always reliant on the social context to maintain. The point is that meaning is always in flux. However, the implications of the contingency of meaning are perhaps more profound: meaning is always political, with uneven social and economic impacts in the way that it comes to shape practices.

As I will explore in more detail soon, the assumption of the 'local' informs how adaptation is applied, and excludes (at present) mobility as an adaptive response to climate change. This is not explicit in the aforementioned UNFCCC definition, yet I will argue that it is an implicit and rather pervasive criterion. As a result of this, our working notion of adaptation conforms to pre-existing political discourses concerning the relationship between place and identity. The impacts of this affect different people according, often, to their geographical location, wealth, social status and so forth. Within our current political context, the 'local' can be glossed to mean both the immediate locality of the village, town or city, as well as the wider national environment (Xenos 1993). Given that we are living in a world where human mobility is deemed appropriate only when it is tightly controlled, we may very well introduce the erased language of the local and understand adaptation to be the:

> adjustments in **the local** ecological, social or economic systems in response to actual or expected climatic stimuli and their effects or impacts. It refers to changes in **the local** processes, practices, and structures to moderate potential dangers or to benefit from opportunities associated with climate change **within the context of one's community and/or nation-state**. (UNFCCC n.d.; my additions in bold)

If we turn our attention to the importance of climate mobility issues, we are confronted with the substantive influence discursive meaning has on human lives. The current adaptation framework is limited for those who have lost, or will lose, their home to sea level rise, chronic drought, cyclonic activity or any other climate disaster, and find themselves unable to safely return to their dwelling (as material shelter, social network and cultural community). If one's land is lost, destroyed or no longer able to sustain human settlement, the first question that arises for the victim is: 'Where can I go now that I have lost my home?'

The phenomenon of where migrating or dispossessed peoples should go is not a new one. The question of 'where?' is, according to Dufourmantelle and Derrida, 'the question of man' (2000: 52). Where do you belong? Where do you come from? Where will you go? The problem of the displaced raises the challenge of offering hospitality to the victim. Dufourmantelle and Derrida offer a provocative philosophical dialogue on hospitality that sheds light on the contentious nature of this ethical exchange. By hospitality, they are referring to the relationship between host and guest and all the dynamics of welcome and refusal encompassed by it. Hospitality is a profoundly human and distinctly cultural practice. Dufourmantelle interprets Derrida's work on the question of Being (that is, what it means to be human) as centrally concerned with the issue of 'where' one can be. 'Where can I go?' Dufourmantelle argues that the question of 'where?' is one that is 'addressed to a man on the move' (2000: 52). The person on the move is one who is caught between where she was and where she may come to be. 'Where?' invites us to face the tension at the 'origin' of human existence. The strong pull of the ontopological, of origins, meets the equally powerful force of displacement as a primary condition. Displacement is 'no less arch-originary' as nativity, a theme I return to over and again throughout this book (Derrida 2006: 103).[4] The foreigner question, 'where?', haunts the ontopological.

Where can I go now that I have lost my home? Contemporary adaptation policy and discourse cannot adequately respond to this question. Yet the question demands our attention. Thus, the question of 'where?' is not naïve. Immediately we might assume that adaptation should occur in one's homeland or recognised (that is, legally prescribed) place of belonging and/or residence. This is why the question 'where?' is so troublesome, despite its ancient traces in all human cultures. American political theorist Thomas Nail contends that the assumption of the local is the result of a dominant political theory that elevates the importance of stasis and the nation-state. Our Enlightenment tradition of political thinking has always returned to a grounding that links the state and the citizen. At the same time as privileging the citizen, political theory has simultaneously relegated movement and migration to the margins of political thought, marking them as secondary considerations (Nail 2015).

By illuminating the coupling of politics–citizen as the unquestioned assumption of politics, we can better understand how it comes to be that adaptation is associated with stasis, or staying-put. National adaptation plans (NAPs) emphasise 'original settlements' and commit the relevant state authorities to efforts aimed at reducing the need for migration or relocation elsewhere (Martin 2014: 20). For instance, Kenya's NAP aims to 'highlight the importance of adaptation and resilience building actions in development' (Ministry of Environment and Natural Resources 2016: 3), while the Nepalese government has retained a focus on development items such as agriculture, water, and food security; all locally significant issues (Ministry of Environment 2010). As I will demonstrate, this political commitment to stasis informs the resolution to mould adaptation planning into pre-existing development agendas. In practice, this means that climate adaptation is encompassed under the umbrella of broader development goals. In effect, adaptation is understood as taking place *in situ* (locally). An *in situ* climate adaptation–development nexus emerges. Development agencies are thus charged with the on-the-ground work of ensuring local adaptive settlement.

THE CLIMATE–DEVELOPMENT AGENDA

Given the priority for the original settlements in current understandings of adaptation, it makes practical sense to subsume climate change issues under the affairs of the international development sector. International efforts to manage the impacts of climate change on humans are reflected in the 'climate change turn in development work' (Bose 2016: 168). Financing is aligned with pre-existing development networks and aimed at promoting local solutions. Simply, international financing is limited to *in situ* adaptation. Wealthy countries, such as Australia, voluntarily contribute economic aid which is then tendered to various NGOs. This is part of fulfilling pledges made to various climate funds, which are loosely guided by CBDR principles. For instance, Australia has pledged $200 million to the GCF. At the time of writing, the Department of Foreign Affairs and Trade (DFAT) was responsible for the distribution of Australian foreign aid. DFAT advertised its role in the following manner:

> Australia is working to find practical solutions to the climate challenge. We are working with countries in the region [Asia Pacific] to manage natural resources, respond to the impacts of climate change and natural disasters and find pathways to sustainable economic growth. (DFAT 2016)

Note here the emphasis on the 'management' of natural resources and commitment to economic growth, both companions of the prevailing global neoliberalism; a condition cultural theorist Henry Giroux argues is reflective of a faithful commitment to the 'belief that the market should be the organising principle for all political, social and economic decisions' (2008: 1). As a set of practices exceeding the realm of the strictly economic, neoliberalism is often bound up with liberal democratic ideals of governance. Leigh Glover explains:

> Actions by states striving to protect existing economic and social conditions from future change and have sought solutions according with the prevailing social orthodoxy of liberal democracy, by applying technological solutions, managerial approaches that combine nation-state powers and market-based policy mechanisms. (Glover 2006: 168)

Given the reliance on international aid provided by national governments, the development sector cannot escape this influence. In the 1980s and 1990s, international aid was explicitly neoliberal in its agenda, positing that economic development would eliminate poverty. Eventually, this was recognised to be a delusion. From the mid-1990s, the sector has shifted away from such a severe adherence to neoliberal ideals and towards a mixed approach. A state–market model is now operating, with a return to concerns regarding poverty reduction in accordance with the Millennium Development Goals. However, according to development theorists Murray and Overton, policies characteristic of a mixed model, such as public–private partnerships and state–civil alliances, 'are in fact ways of legitimising and sustaining the dominant regime of accumulation' (2011: 317).

Of the $200 million Australia pledged to the GCF, $34 million has been used to fund 'Community-based Climate Change Action Grants' aimed at addressing 'development needs at the community level' (DFAT 2016). Over an eight-year period (2008–16), $12 million was channelled into the United Nations Development Programme initiative 'Small Island Developing States Community-based Adaptation'. This project focuses on water supply, coastal management and agriculture within communities across thirteen Pacific states; valuable and needed work, no doubt. Yet we should heed development economist Erik Reinert's warning that, informed by neoliberal targets of economic growth, even well-meaning development interventions are akin to introducing the palliative care team in at the last minute to assist someone in a dire situation to feel more comfortable (2007: 240).

Notwithstanding Reinert's critical view, it is useful to 'consider the likelihood, strength, and interaction of climate change impacts' in development strategies (Müller et al. 2014: 2505). To do so is to align with the interests of the community in preserving and strengthening their relationship to land and culture. However, at present it is unclear where adaptation aid goes, and if it is actually used to develop strategies the local communities regard as important (Betzold 2015: 487). Technological 'solutions' that donor countries may consider concrete 'outcomes', such as sea walls, are often 'unsustainable and ineffective against shoreline erosion . . . built without consideration of the specific conditions. . . . Seawalls thus collapse, leaving the community as, or even more, vulnerable' (2015: 487).

In the Pacific, Small Island Developing States are especially vulnerable to the impacts of climate change. Australia's financing schemes are therefore welcome and necessary, if limited. Numerous countries are at risk of complete destruction due to rising sea levels. For instance, Tuvalu, a country of roughly 12,000 people, is only 5 m above sea level at its highest point. Since 1993, satellite data indicates that the local sea level has risen by 5 mm per year, a figure higher than the global average (Pacific-Australia Climate Change Science and Adaptation Planning Program 2016). The IPCC predicts that sea levels will continue to rise at 4 mm on average per annum (2013). It is anticipated that this will leave island states such as Tuvalu uninhabitable by 2100.

Nearby, the nation of the Marshall Islands is only 3 m above sea level at its highest peak. With 55,000 residents, its president Christopher Loeak has stated that 'the Pacific is fighting for its survival' (Vidal 2013). Similarly, Kiribati's president Anote Tong has warned that his land will likely become uninhabitable by the end of the century. In the face of these comments, international mitigation and adaptation action seems to fall far short of what is needed. Sea level rise and threats of land loss exceed the operative concerns of development. The question 'where?' is brought into clear focus (that is, where should these people live?). Just as Dufourmantelle and Derrida argue, this question reveals the primacy of our relationship to place and the imminent possibility of placelessness. Why? Because our neighbours may lose their lands and this in turn puts into question our entitlement to specific places. As I will develop in the following chapters, the question 'where?' confronts all of us to reflect upon our own relationship to place. Our land and our place are inherently unstable. There is a similarity in conditions between 'their' situation and 'ours'. This connection needs to be cultivated in order to spur ethical action that sustains their lives as well as our own. This is the ethical challenge climate change presents us with.

I have painted a frightening picture of the potential impacts of climate change in the Pacific. Human geographer Carol Farbotko rightly notes that perpetuating narratives of drowning Pacific Islanders is both empirically misleading (at least in the short term) and under-estimates the resilience of locals to adapt to changes locally, without the need to migrate or relocate (2010). However, it is equally important that we question the various motivations of countries, such as Australia, in supporting the mainstreaming of climate adaptation into development to the exclusion of, at present, migration and relocation options. The Australian government has also collapsed adaptation funding into foreign aid, as well as reducing its contribution to international aid to below the global minimum of 0.7 per cent of Gross National Income (Wade 2016). Foreign aid is exactly that: it seeks to assist foreigners in their home country. The Australian government has made its financial contribution to adaptation funds conditional upon this. In 2007, a call by the Australian Greens for the recognition of climate migrants was rejected by the Australian government.[5] As I will discuss in greater detail in Chapter 3, when the international Advisory Group on Climate Change and Human Mobility sought to elevate the significance of migration as a form of adaptation at the Paris Conference in 2015, Australia was quick to reject the proposal, successfully removing the agenda item from the negotiating table.

The Australian government's refusal to entertain the possibility of mainstreaming mobility options in the Pacific is motived by more than a genuine desire to preserve local cultures. In the Pacific, Australia's contribution to the GCF and its role in resilience building, that is *in situ* adaptation, is unfolding in a political context hostile to specific types of immigration. Taking a longer, historical approach, migration between Australia and the Pacific Islands has been subject to racially motivated restrictive practices. Alongside the passage of the *Immigration Restriction Act 1901* in a newly federated Australia was the *Pacific Islander Labourers Act 1901*. The latter was a law which authorised the deportation of all Islanders residing in Australia at the time, and mandated the condition that entry into the country could only take place as a form of slavery called indentured labour (Docker 1970).

In a contemporary setting, Australia's relationship with various Pacific countries, Papua New Guinea and Nauru in particular, is embroiled in the violent incarceration of asylum seekers who have attempted to make their way by boat to Australia and, with this, the proliferation of ad hoc and dishevelled prison camps (Gleeson 2016). Strategies of migration control extend well beyond the borders of the sovereign state, signalling the consolidation of a migration management system that is 'delocalised' (Pécoud and de Guchteneire 2006: 71; see also Taran 2001; Vaughan-Williams 2015; Mountz 2011). The unequal relationship between Australia and the Pacific is compounded by the power imbalance in the region, with Australia providing more than half of the international aid offered to the Pacific region; aid which many Pacific countries are dependent on economically. In Reinert's (2007) view, when development income makes up such a significant proportion of a country's domestic budget, it produces a form of 'welfare colonialism' which, rather than empowering locals, risks encouraging passivity.

The climate–development agenda is not politically or economically neutral. The brief example I have outlined here illustrates the complex power relations that precede the problems of climate change yet shape how this issue is addressed. Geopolitical relations between Pacific states and Australia bring to the fore the manner in which climate change adaptation intersects with pre-existing and contested issues concerning mobility and place.

ADAPTATION: A POLITICS OF PLACE

The alliance of climate adaptation with development platforms reflects back to us the reality that climate policy contains within it a series of underlying political imperatives concerning geographical places. Most importantly, the nation-state retains its control over matters related to land, resources and immigration. The UNFCCC 1992 repeatedly makes clear that climate change does not challenge the authority of the sovereign state to make decisions about its territory or peoples (and by extension, the arrival of foreign persons into national territory). At one point, quite early in the Convention, the 'principle of sovereignty' is 'reaffirmed' (1992: 2).

In other words, adaptation involves the negotiation of issues associated with the right to dwell in a particular place. Political theorist Andrew Kythreotis reminds us that territory is a core part of climate politics when he argues that '"territoriality," both as a material and discursive device, is fundamental in, and constitutive of, how we interpret and understand climate change and the politics thereof' (2012: 457). What he means by this is that how land (and resources) are marked and defined as territories belonging to or excluding particular entities (states or corporations or persons), is a core consideration in our climate change discourse. Climate adaptation is primarily about place: it is about how we value place, whose place matters, who has the right to a place, as well as what our duties are towards our own place and that of others. Today the organisation of the political community into the model of the nation-state means that place is subject to the logic of 'territorialisation'. Territorialisation refers to the process of dividing land into parcels that are contained by clear boundaries, confirming anthropologist Liisa Malkki's point that in spite of the powerful forces of globalisation, 'boundaries and borderlines are at the centre of our analytical frameworks' (1992: 25). The issue that most fiercely and consistently challenges this organisation of land into discrete parcels is human mobility.

Malkki, writing on the dislocation and resettlement of refugees in the 1990s, has argued that the nation-state assumes a metaphysical status in relation to the management of mobile peoples. This metaphysical status emerges in the way that the nation-state dominates how we politically and legally organise homeliness and belonging to place, in turn generating a discourse of national roots: rootedness in place is then nationalist (Malkki 1992; see also Xenos 1993). Perhaps ironically, this desire for roots occurs in a context in which 'people are chronically mobile and routinely displaced' (Malkki 1992: 24).

Malkki takes her cue from the philosophical work of Deleuze and Guattari who argue that belonging and identity are not fixed points and instead take on more rhizomatic forms. Rhizomatic identity cuts away at a linear and singular notion of identity, the latter likened to the arborescent structure of a tree root system. The metaphor of the rhizome suggests that our identity is always networked and messy. In this way, we can view identity like we might see the root system of a patch of grass: criss-crossing, complex and without a single point of origin or end. If we follow Deleuze and Guattari's line of thinking regarding rhizomes, we can understand that mobile populations continue to make meaning in spite of the absence of a territorial origin (Malkki 1992: 24). That is, rather than anchoring identity in a singular place of origin, an ontopological assignation, identity forms intricate and non-teleological networks of connection.

Such flexibility or complexity between people and place is exemplified in the Pacific where dwelling and mobility are both integral to culture, as I will discuss in due course. Yet despite evidence of rhizomatic identity formations at the grassroots level, Malkki contends that the dominance of 'a peculiar sedentarism' is 'reflected in language and in social practice' regarding the resettlement and rehabilitation of refugee communities (1992: 31). Working through development agendas, this 'peculiar sedentarism' is replicated in current adaptation funding and implementation models. Faithful adherence to sedentarism is 'peculiar' because it appears to be blind to the historical condition of human migrations as norm rather than aberration. However, it is less peculiar when we consider how a belief in settled communities perpetuates the idea that outsiders are threatening and invasive.

External threats to national territory play both a positive and a negative role in motivating the global community to create structures for the management and containment of peoples and lands. In a way, adaptation discourse does address the issue of migration, but as something that is deeply threatening to the dominant political organisation of place as national territory. This is not entirely problematic. The adaptation agenda attempts to enable communities to remain on the lands that constitute their place of belonging, their homes, or what political theorist Hannah Arendt refers to as their 'social textures': the web of relationships and structures that build and sustain social being (1951: 372). It is the ethical and social role of development organisations to facilitate community building and resilience: to foster the capacity to be 'safe and in place' in a sustainable manner (Martin et al. 2013). However, the emphasis by governments and development agencies on sustaining local livelihoods and homes tends to marginalise the reality of widespread chronic displacement and the increasingly important role of migration and even relocation in adapting to climate change. Indeed, the IPCC *Fifth Assessment Report on Climate Change* recognises that forced migrations will 'compromise human security' and that mobility will be 'a widely used strategy to maintain livelihoods' (Adger et al. 2014: 758).

In other words, the capacity to adapt locally is affected by economic, environmental, social, political and psychological factors that may exceed the remedial capacities of many development programmes. For instance, in Bangladesh, 63 per cent of household incomes are dependent upon employment in the agricultural sectors (Kartiki 2011: 25). Should drought, severe weather or chronic flooding impede efforts to cultivate land or conduct fishing activities, primary livelihoods are dramatically affected. Katha Kartiki provides an extensive study of the impacts of repeated exposure to severe climate events for coastal communities in Bangladesh, writing:

> Coastal villages are typically densely populated and regularly experience coastal floods, river erosion, saltwater intrusion and other natural calamities made worse by sea-level rise. Increasing sea-level in the future, coupled with existing problems, can push many on the move. For instance, in 2007, this area was hit by cyclone Sidr, a category 5 cyclone. This was followed by a category 1 cyclone, Aila, in 2009. This caused widespread deaths, large-scale destruction of livelihoods and property forcing villagers to migrate. As repeated exposure to such disasters depletes the asset base of households, it makes future recovery very hard. (Kartiki 2011: 25)

While many would prefer to remain in their homes and with their communities, it is increasingly difficult. One village elder conveyed with despair that 'there is nothing to do but find another home', an option referred to as the 'last survival strategy' yet one that is taken up with greater frequency. Kartiki points to the disproportionate percentage of people 'forced to migrate after [Cyclone] Aila [who] could not find money to enable them to pay for the cost of migration', a point which foregrounds the need to expand financing schemes and adaptation plans to include local migration support as a social justice imperative (Kartiki 2011: 28; Siddiqui 2010; McAdam and Saul 2010; Martin 2012). Yet sedentarism prevails in adaptation policy.

Bangladesh illustrates the challenges of local adaptation as families and individuals struggle to maintain their networks and livelihoods in environments that are unstable and subject to frequent and turbulent transformations. While it is the security of the persons themselves that is dangerously at stake, in the broader political and social setting, human mobility itself is conceptualised as a problem, a threat, a cause of insecurity for the international state system. Mobility is not seen as a rather reasonable outcome, and response, to volatile political and ecological contexts but as an 'unfortunate phenomenon' (Nail 2015: 26).

It is incontestable that adaptation has undergone important conceptual changes with regards to the proposed 'agent' of change. A biosocial understanding is cognisant of the pivotal role of humans in both climate change and how we respond to it. This is not the end of the conceptual story though. The shape that adaptation takes in practice is subject to historical conditions and political forces. Adaptation governance is organised within a political and economic context in which nation-states continue to define the categories of belonging and exclusion, as well as responsibility and authority; reasserting old ideas of blood and soil or belonging and birth-right.

THE ETHICAL PRIMACY OF THE LOCAL

The above analysis is concerned with articulating the political construction of adaptation. Pivotal to this is the privileged site of the 'local'. The construction of the 'local' is a political process that veils over or marginalises three important factors. Firstly, the complex relationship many people have to place (Jansen and Löfving 2009).[6] Secondly, focusing on the local tends to ignore the demographic reality of widespread, if highly variable, migration. Lastly, when we limit ourselves to local considerations we marginalise the rights and needs of mobile peoples, whether displaced, migrating or relocating. These three reasons, alone, are compelling enough to emphasise the need to engage more seriously with the political and ethical challenges of climate mobility.

However, it is imperative that we hold our critical lens together with an appreciation of the ethical contours of any appeal to local adaptation. There are several significant reasons why the demand for local adaptation has been taken up. As I have stated already, adaptation is primarily about our dwelling; it is concerned with our 'place' in the world. Home and dwelling are intensely cultural concepts, thus calls for local adaptation cannot be reduced solely to the interests nation-states have in containing populations. Community, identity and belonging are at stake. The ethical significance of the local place can be appreciated by focusing on two themes: 'cultural roots' and 'rights discourses'.

The notion of cultural roots can provide an important impetus for the empowerment of disadvantaged groups throughout the world. Similarly, rights have been taken up in ways that attempt to promote cultural and communal preservation. Roots and rights wrap themselves around broader calls for intergenerational justice and distributive responsibility. Anchoring identity in cultural tradition provides a form of intergenerational justice embedded in the customs and beliefs of the everyday. Roots aim to preserve the past for the future. In addition, through development programmes, informed as they are by a rights paradigm (Uvin 2007), there is an implicit, though weakly implemented, commitment to distributive responsibility in the sense that foreign aid is often North to South and aims to redress global inequalities resulting from the uneven spread and impact of free-market capitalism.

'Ethical' Roots

Though dated, Malkki's work from the early 1990s exemplifies a theoretical shift away from fixity and origin towards fluidity and movement. The idea that the nation-state, or even one's hometown, offers a sense of primal belonging, sits uneasily within postmodern and poststructuralist perspectives emerging since the 1960s. Today, movement is viewed as characteristic of social life and individual experience with the dominance of global flows of people and goods (Ahmed et al. 2003: 2; Nail 2015; Urry 2007). Nonetheless, Malkki's critical lens is useful for two things. In the first instance it challenges the assumption of a naturalised and innately desired relationship between blood and soil.[7] In the second instance, shining a light on the sedentary politics of the nation-state reveals the investment these institutions have in the maintenance of sovereign borders and the politics of belonging and exclusion.

Nonetheless roots, blood and origins continue to carry personal salience, as well as cultural import, and consequently impact the formation of government and non-government intervention into political issues, including climate change. It may be tempting to simply apply Malkki's thinking to climate adaptation discourse, but such a reading would obscure the complexity of the social worlds affected by wild and unpredictable weather. We need, here, to disentangle sedentarism and roots. Sedentary politics, when carried out at the national level, reproduces nationalist roots, but roots are not reducible to expressions of nationalism. The logic of roots may be viewed as a powerful act of resistance to the forces of climate destruction, neoliberal economic expansion and global political negligence. In the Pacific, for instance, Kiribati's residents and leaders continue to fight for international action on climate change in an effort to reduce the need for migration or resettlement elsewhere. Local activist Maria Tiimon has urged us to question the assumption that all disadvantaged peoples wish to move:

> I speak to the young people there and they say they don't want to move. This is where our ancestors came from. . . . Displacement really has to be the last resort. Pacific islands need help to adapt and the rich countries need to cut their greenhouse gas emissions. (Milman 2015)

The land holds tradition and identity. These cultural roots are threatened by forced displacement.

However, when we urge others to accept the validity of our cultural roots we are not necessarily ruling out the role of migration as a culturally appropriate response to climate change. Tiimon's call for the recognition of cultural rights may emphasise the physical presence of land but, as Lynette Carter has written, 'to maintain people's identity they do not have to be physically *locked into* specific places' (2014: 59). Note the language here: while place is a site of significance, identity is mutable and mobile; strong enough to move between places, highlighting, perhaps, Malkki's point regarding identity in mobile communities. This dynamic is difficult to accommodate within a state-based global system. In the nineteenth and twentieth centuries, the colonial imposition of stasis oppressed, in anthropologist Epeli Haú ofa's view, the historical complexity of mobility and dwelling in the Pacific:

> Human nature demands space for free movement, and the larger the space the better it is for people. Islanders have broken out of their confinement, are moving around and away from their homelands, not so much because their countries are poor, but because they had been unnaturally confined and severed from much of their traditional sources of wealth, and because it is in their blood to be mobile. (Haú ofa in Carter 2014: 65)

In the Pacific, mobility is part of cultural identity alongside strong cultural relationships to land and place. Place, identity, belonging are not as linear as the notion of 'blood and soil' or birth–nation–belonging. When this cultural difference is transferred onto the debates concerning adaptation, the concept of 'local' is placed under contention. That is, a call for the preservation of cultural roots and ancestral lands can co-exist with the uptake of migration by locals. Our adaptation projects must hold both, thus the aforementioned focus contained in Australian adaptation aid is limited and only partially responsive to cultural context.

Local adaptation can be seen as a set of practices aimed at the pragmatic concern for current and future habitation and existence. More than this, however, it is a process that implicates issues of cultural and social (and often spiritual) belonging. As evident in the Pacific context, this is not the same as claiming an essentialised relationship between identity and land. Instead, what is important is the relational and dynamic connection between people and the places they live. We hear in Haú ofa's words the effort Pacific Islanders have had to expend in order to 'break out of their confinement'; the colonial imposition of discrete territorial configurations, which has oppressed the energy of 'free movement' found 'in their blood'.

We can view the call for cultural roots as distinct from the territorial logic of ownership-over. Unpacking the way that territory semantically operates, loosens the conceptual strictures of 'home'. Roots are not exclusively nationalist and can be articulated powerfully as part of a local empowerment discourse for marginalised peoples and communities. There can be an ethical acceptance of a discourse of roots to support *in situ* adaptation action without the simultaneous need to exclude mobility options.

An 'Adequate' Adaptation

Cultural roots are one way to provide recognition and importance to the relationship marginalised and/or colonised peoples have to their lands. Communal and individual agency is enabled. From this position disadvantaged communities can intervene in the political construction of adaptation, highlighting the ethical significance of cultural and spiritual belonging alongside the dominant focus on economic interests.

The call for rights recognition is another way local communities participate in the political construction of adaptation. Since the 1990s, development organisations have increasingly utilised rights-based frameworks in their platforms for action. As Peter Uvin points out, this shift in rhetoric signalled a rearticulation of the role and agenda of development organisations from 'service based' to 'rights based' (2007: 600). Interventions had to be more than just about need or survival. Traditional empowerment, advocacy and participatory discourses gelled with rights approaches. Now, however, an emphasis was placed on the responsibilities of the relevant nation-state towards its citizens:

> To have a right to something, say food, is not just about having enough of that: a slave can be well nourished too. It is about having a 'social guarantee' . . ., which implies that it is about the way the interactions between citizens, states, and corporations are structured, and how they affect the most marginal and weakest in society. (Uvin 2007: 600)

How we conceive of and create 'home' is intimately bound up in this political process of negotiating universal rights regimes within specific cultural contexts. These frameworks articulate the foundational rights individuals and communities have to retain and develop an 'adequate' livelihood and sense of homeliness; the 'social guarantee' of security or safety. If implemented well, rights-based adaptation strategies enable human flourishing and not mere survival.

Home does not begin and end with the structures of a roof and four walls. The responsibility to create the conditions for homeliness extends from the material or physical to the cultural and the social. In international doctrines the notion of 'adequacy' appears repeatedly, encapsulating a diverse grouping of social, political, cultural and economic concerns that amount to the 'social guarantee' of human dignity.

Article 25 of the *Universal Declaration of Human Rights* 1948 (UDHR), for instance, asserts that:

> Everyone has the right to a standard of living adequate for the health and wellbeing of himself and of his family, including food, clothing and medical care and necessary social services, and the right to security in the event of unemployment, sickness, disability, widowhood, old age or other lack of livelihood circumstances beyond his control.

The *International Covenant on Economic, Social and Cultural Rights* 1966 reiterates the definition of 'adequacy', but emphasises the role of the state in taking 'appropriate steps to ensure the realisation of this right'. The elaboration of social and economic rights in the *1966 Covenant* highlights a shift from a more 'classic' paradigm of individual rights, as per the UDHR, to a 'welfare rights' emphasis (Vincent 2010: 117–23). With this, significant disagreements arose as the ideological tensions of the Cold War informed how different national governments engaged with the newer rights paradigms, with some Western states expressing resistance to embracing what were perceived to be socialist ideas, such as welfare.[8] Rights associated with housing are heavily informed by these ideals. For instance, Article 11(1) of the *1966 Covenant* highlights the point that appreciation of adequate housing extends from an 'adequate standard of living' to the provision of 'food, clothing, housing and the continuous improvement of living conditions'. The inclusion of 'continuous improvement' points to the inadequacy of treating 'home' as merely shelter or a physical structure. Such physical structures are only purposeful and equitable if part of a broader network of social services and opportunities and ensured by the state.

More recently still, it has been acknowledged that the conditions of adequacy must encompass the concept of a healthy environment able to sustain human life into the future. Since the early 1970s it has been understood that human rights are reliant upon a healthy environment, and indeed the 'right to a healthy environment' must be expressed alongside other human rights. The *Stockholm Declaration* 1972 was the first written acknowledgement of this interrelation:

> Man has the fundamental right to freedom, equality and adequate conditions of life, in an environment of a quality that permits a

life of dignity and wellbeing, and he bears a solemn responsibility to protect and improve the environment for present and future generations.

While these are ambitious proposals, they are non-binding and as such the implementation in domestic legislation has been haphazard and slow. 'Continuous improvement' may point to the need to ensure that the living situation is sustainable in the long term, an issue relevant to the problems of climate change, which alter the dynamics of sustainability at the fundamental level of geography. However, it is important to note that in the Pacific region, Kiribati, Tuvalu, Nauru and the Marshall Islands are amongst those that have failed to sign or ratify the *1966 Covenant*. Similarly, these nation-states are not parties to the companion covenant on civil and political rights, the *International Covenant on Civil and Political Rights* 1966 (see Farran 2007).

Home and Political Participation

The security of a home increases the capacity of people to engage in the local political process. Arendt has called our attention to the link between of having a 'place' in the world, that is, a secure home accompanied by a sense of belonging, with the capacity to exercise political autonomy (1951: 296). In *The Human Condition*, action is regarded as the highest domain of human affairs. Arendt relates the concept of a 'world', and our investment in it, as a hallmark feature of political action (1958). Arendt's concept of 'world' emphasises the importance of intervening in affairs that exceed the everyday needs of survival and instead leave traces for future generations; the space of politics. The ability to create, sustain and transform the 'world' in innovative ways is a uniquely human affair and the highest achievement of politics (Arendt 1958). From this it is not unreasonable to suggest that the ramifications of the loss of home are not limited to forms of social, cultural and economic loss. The loss of 'home' can actually result in the withdrawal from the realm of politics and the capacity to shape the future.

Today, political participation is frequently tied to being able to legally identify yourself and your permanent place of residence. In their report, *Land in the Struggle for Social Justice*, Gelbspan and Thea draw attention to the relationship between security of place, or the availability of affordable housing, and political participation. The authors note the centrality of land to securing human rights as well as the import of a human rights paradigm when negotiating land struggles.[9] Land is vital in urban contexts where levels of housing insecurity are rife, in both parts of the Global South and Global North (Gelbspan and Thea 2013: 21). The relationship between political inclusion and security of place becomes an incredibly fragile and complicated issue when dealing with internal displacement, or internally displaced persons, as well as the expansion of informal settlements. While internal displacement and/ or 'slum dwelling' may not lead to the suspension of rights to citizenship in the formal sense, the extreme marginalisation experienced by the homeless and poor may lead to informal disenfranchisement. This is particularly the case if a fixed residential address is a requirement for voting registration.

For Arendt, political marginalisation, combined with the loss of a social texture, is equivalent to stripping one of his or her humanity and of denying or taking away the very core of being human. A disregard for cultivating the conditions of human dignity still occurs in places where a 'social texture' is present (that is, an informal settlement) but where political forces and relations ensure a lack of security and political inclusion. Oftentimes, these social textures are transient, facing the ever-present threat of sudden and forced eviction, radical exposure to the climate, and dispossession for development purposes. In other words, the presence of a social texture in marginal but resilient communities does not map onto the presence of political rights and protections. When one is excluded from the privileges of political participation, this often leads to difficulties retaining and sustaining a home, or securing a new home, in the event of its loss. Thus, an adequate adaptation must include within it the assurance that community members can actively participate in the political processes of decision-making and change for the future.

Rights instruments and development programmes try to contain the dynamism of the home; every new protocol or right aims to make stable and equitable that which is inherently contingent due to, at an empirical level, its reliance on external factors, including the environment. Climate adaptation, when paired with a development ethos, works to provide a platform for creating protection and human flourishing in conditions that frequently make this difficult to attain and sustain. However, because development programmes are indebted to state financing and subject to the imperatives of global neoliberalism, action on climate change can be subsumed by the requirement to continue to 'grow' economically. At the macro level, an ethos of distributive justice and intergenerational responsibility informs international decision-making. However, these norms and ideals operate in contexts where states may not be signatories, and even if they are, the primacy of the sovereign retains critical purchase.

CONCLUSION: ON SHAKY GROUNDS

It is becoming increasingly apparent that our conceptual and concrete expression of adaptation needs to expand beyond assisting safety-in-place and look to how we can respond ethically to displaced populations, mobile peoples, and communities facing permanent relocation across international borders. Just as what constitutes adaptation is politically constructed, its conceptual and practical limits are the outcome of power struggles. The limits that I am pointing to are not ecological in nature; yes, the climate may provoke the need for action but the solutions are not determined solely by environmental limits. Rather, they are limits imposed by political imperatives, social values and the diminished import of ethics in practice. As Adger et al. have argued, adaptation is a set of contingent, rather than fixed, practices:

> limits are endogenous and emerge from 'inside' society. In this reading, what is or is not a limit to adaptation becomes a contingent question. It all depends on goals, values, risk and social choice. These limits to adaptation are mutable, subjective and socially constructed. How limits to adaptation become constructed rather than how they are discovered, becomes the operative question. (Adger et al. 2009: 338)

The evolutionary logic of adaptation as biologically driven has given way to a more nuanced understanding of the social forces that affect our capacity to both survive and sustain a healthy relationship to place. Adaptation-as-survival is a political issue which highlights the profound inequalities that structure our global governance system. Simply, human life is differentially valued. Through the application of normative paradigms such as distributive responsibility and intergenerational justice, dominant discourses of 'climate ethics' aim to contest the injustices that mark the impacts of climate change. Within this framework, responsibility requires the calculation of costs and benefits or 'egalitarian redistribution' (Parr 2012: 18). Adaptation financing is subsequently funnelled into pre-existing development pathways. From here, counter-discourses of adaptation emerge. Adaptation is not merely survival but communal flourishing: local empowerment draws on the language of roots and rights to promote a narrative of adaptation as a set of resilience practices.

However, the challenge of climate change illustrates the significance of the relationship between identity and place as an ongoing and contested political and ethical concern; an invitation to reflect on our right to place or, more fundamentally still, to be at-home in the first instance. Indeed, we are living in an exceptionally stressed political arena subject to ecological changes that are literally threatening the very capacity to demarcate territory. Demarcation is the effort of ensuring permanence in a world in flux: land is parcelled up and the bodies moving across it are regulated and identified according to whether they 'belong' to that place or not; an unsustainable ontopology. Human displacement, migration and relocation are already conditions that provoke an aggressive form of territorial protection. Dangerous climate change adds to this scenario the inevitability of complete territory loss. Climate change threatens the capacity of states to govern and control both human movement and geographical territory. In this way, it confounds our contemporary political logic, predicated as it is on the presumption of stasis. Perhaps more radically, climate change challenges us to look honestly at the groundless nature of our existence, and its transitory and contingent expression; the impossibility of being at-home. The ground, after all, is literally moving under our collective feet.

Chapter 2

RETHINKING ADAPTATION: AN ETHOS OF DWELLING

'I cannot disentangle myself from society with the Other.'
Emmanuel Levinas[1]

'Climate change is gradually divorcing us from our land and eroding our subsistence way of life. Please think for a moment how you would react if climate change threatened your very existence as a distinct people.'
Inuit activist Sheila Watt-Cloutier[2]

The crisis of human-induced climate change brings us back home. It is an issue which invokes the fear of species annihilation, certainly, but more immediately draws us into confrontation with the way in which we inhabit this earth. In mainstream climate change literature, adaptation is taken to refer to the adjustments, pre-emptive or reactive, that people make in response to predicted or actual climate events. These alterations can occur at the level of social, economic or ecological processes, practices and structures. In other words, at present, our adaptation discourse posits that with greater economic resources and more advanced technology we will limit the impacts of climate change and assist with any adaptations as necessary (Glover 2006).

This reduction of adaptation to a series of 'technical problems' (Glover 2006: 170) – which require cost–benefit analysis (a neoliberal logic) – fails to appreciate the deeply philosophical question inherent in the desire to adapt. Adaptation is about future habitation. What else is survival for but the promise of a future? The failure to survive – death or extinction – is the antithesis of adaptation and dwelling. This concern for future habitation reveals to us the fact that adaptation is centrally about 'dwelling'. For Heidegger, authentic dwelling confronts mortality as a condition of existence and works towards a 'good death' (1978a). In Levinas, the possible death of the Other contests my claim to dwell without interruption; I must welcome the Other and renounce the violence of my possession. From this perspective, adaptation is a political contestation over the foundations of the dwelling place that is both preceded *and* exceeded by ethics. This concern for the future is ultimately ethics. But as Derrida writes, such an obligation cannot be postponed as 'this future, this beyond, is not another time, a day after history. It is present at the heart of experience' (1978: 118–19).

In Chapter 1, I explored the manner in which adaptation is politically constructed. Adaptation is not only a set of actions geared at sustaining life or minimising risk. I have delved into the primary issues of place and identity at stake when we begin to talk about adaptation to climate change; its ontopological commitments. Normative ethical approaches such as distributive responsibility have loosely informed some of the political decisions related to adaptation. Inadvertently these ethical models tend to reinforce a separation between self and other, and local and foreign. The way that we respond to the challenges of climate change cannot rest solely on the application of rational principles. We need an ethics that is grounded in the relational and responsive to the needs of the Other. Adaptation is mixed up with the messiness of human co-existence. It is because of this that it produces such unresolvable yet urgent ethical dilemmas. And it is precisely in the promise of connection contained in such human messiness that the potential arises for forms of ethical action attuned to relationality, interdependency and an acceptance of groundlessness.

In this chapter, I put forward a relational ethics as the philosophical basis for appropriate climate action. Rather than calculate and measure, or determine a nation-state's 'index of responsibility' (Baer 2010: 252), I develop an ethos of dwelling grounded in care, finitude, inter-subjectivity and the *primacy* of the Other. There is an inter-subjective relation in normative theory too, but it tends to fold back into self-contained individuals interacting with one another. The radical aspects of Levinas and Derrida, as I highlight in this chapter, refuse to settle on this foundation of self-possession or undisturbed 'at-homeness'. Dwelling is never simply an issue for myself or 'I' alone, nor can it be reduced to a set of duties based on cost–benefit analysis or reference to precise degrees of responsibility.

Through a discussion of Heidegger and Levinas, I argue that

to dwell-well is to dwell-with. I will outline the key ideas I take from Heidegger (including 'dwell-with' or *Mitsein*) before developing Levinas's contributions as a more useful ethical model for thinking about responsibility for Others. When we dwell-with we attune ourselves to the immense pain experienced in the event of cultural, spiritual and material loss or damage of one's home and homelands. However, we must go further than just dwelling-with as an attunement: we must dwell-with the suffering of the Other. In turn, we come to understand that another's current or predicted loss calls our 'being at-home' into question and, in turn, this calls upon us to act.[3] For example, the reality of communal relocation as a result of climate change demands that we both recognise and respond to the suffering of the Other. To close this chapter, I explore the ethical implications of the Inuit petition calling for the 'right to an inviolable home' in light of the ideas of Heidegger and Levinas. Who is responsible for the consequences of climate change for aboriginal communities in Alaska? And what form should responsibility take? What should we do?

DWELLING

To get to this relational ethics, we start with 'dwelling'. Dwelling is about how we live independently and with others on this earth. Confronted with the enormity of rebuilding and recovery following WWII, Heidegger asked his fellow citizens to consider the following question: 'What is the state of dwelling in our precarious age?' (1978a: 339). His answer was not encouraging. It staged a critique of the complacency towards existence that comes with the emerging dominance of technologisation in the industrial age. Technologisation refers here to a process of ruthless instrumentalisation where everything is knowable only according to its economic value. For Heidegger, a grave consequence of this is evidenced by the disconnection of modern society from cultural traditions, including the importance of language for issues of identity and belonging, the environment, and the meaning of Being. As I noted in Chapter 1, language is bound up with social power structures. How we give meaning to life through language tells us about our values. Is the earth a 'resource' or a 'home'? Modern technology seeks to reduce everything according to its utility; its capacity to be exploited and used up. For Heidegger, dwelling, thinking and Being are to be considered together. How we dwell is how we understand our place in this world and the very fibre of Being. The task of adaptation compels us to ask similar questions of ourselves.

Dwelling brings us back to what is primary yet has been forgotten, that is, our relationship to the earth; an earth in which we are both 'entrusted and exposed' (Heidegger 1978a: 339). We are entrusted with caring for this world, yet we are also exposed to the conditions of earthly dwelling. That is, we must come to embrace the fact of our finitude and fragility on the earth as well as our unending responsibility to care for this place that will hold us in death. Human geographer Nigel Clark (2011) refers to this condition as one of 'radical dissymmetry' in which we humans are absolutely reliant upon an earth that does not require us to continue with its existence. Yet we also find ourselves in a context of mutual vulnerability with the impacts, or footprints, of human societies causing great destruction and disrepair to the earth's functioning. Indeed, we are, by all credible accounts, ruining this dwelling place.

It is time to consider Heidegger's question once more. Only now we are asking, 'What is the state of adaptation (as dwelling) in our precarious age?'

PLIGHT

When Heidegger urged his fellow citizens to think on the state of dwelling he was specifically asking that his contemporaries recognise the ruinous relationship humanity had developed towards the environment. A lack of concordance with the environment was the biggest reason for our plight; our inability to dwell well. It is the 'will to master', inherent in the endorsement of modern technology, that produces the 'plight of dwelling' (Heidegger 1978b: 293–7). [4] When we seek to dominate we turn away from the embrace of our fragility and mortality, and fail to take care. This is because technologisation produces the illusion of immortality. In this sense it is reckless. For Heidegger 'man in the technological age is, in a particularly striking way, challenged forth into revealing' (1978b: 305). In the Heideggerian lexicon, 'revealing' refers to the process of (often) violent extraction. That is, nature is to be exploited, to be violently extracted, for all its worth and at any cost (1978b: 305).

When something is challenged into revealing itself, it is not permitted what we might call its own agency or cycle. For Heidegger, technology reduced to instrumentality (a means to an end) remains committed to a set of binaries which position technology over nature and subject over object (1978b: 296). Within this binary, a human being assumes the control and mastery of nature. This diminishes our experience of dwelling and our relationship to the earth, ourselves and others, as we too are captured by the violent logic of instrumentality. Fellow humans begin to be viewed in economic terms. The framing questions come to be: what is the economic worth of this? How much economic growth can be predicted to result from this action? Is this land economically viable? Is the acceptance of climate migrants economically beneficial? Is communal relocation 'worth it'? Or, more confronting still, what is the worth of the life of the stranger?

The mantra of economic growth accompanies the dominance of technology as a saviour. We may have become so desensitised to the implications of global neoliberalism that we fail to take in the fact that, as Heideggerian theorist Joronen succinctly puts it, 'the whole of the earth eventually turns into a global resource' (2011: 1127). When this is the dominant attitude, Dasein (Heidegger's term for human being) is uprooted or not at home; this is the plight of dwelling. Stuck in this anxious spiral, we lose ourselves and our sense of purpose. Our days are marked by a 'harassed unrest' (Heidegger 1978a: 328). In the era of climate change, with the domination of the logic of technologisation, we are dangerously lost.

Adaptation discourses have so far been unable engage with loss in this sense: the loss of a 'world'; the profound loss of disconnection. This is the result of a continued emphasis on 'physical' conditions. The environment, built infrastructure, economic processes and technical interventions remain the core focus because they contain an 'analytical functionality' and can provide 'objective measures and thresholds of danger' (Adger et al. 2009: 336–7). They are domains that are readily subject to valuation. However, the plight of dwelling results in devastating loss that exceeds calculation.

The plight of dwelling amounts to more than a philosophical complaint. This plight speaks to the profound experience of loss as well as the intense presence of anxiety and fear regarding the future. What, drawing on Heidegger, have we lost? Nothing short of our capacity to act in the world, to be 'authentic'! According to Mitch Rose, dwelling can be understood as a 'form of practice', 'a response', a 'matter of stance' and a 'matter of acting' (2012: 761–2). But the activity of being-in-the-world can only commence once Dasein has accepted its 'placement in a world that is not its own' (2012: 762). We must embrace our contingency and finitude or, essentially, our groundlessness.

DEATH

The plight of dwelling is secured in the separation we assume from others and the earth. This separation disconnects us from the primary condition of being human as dependent on the earth and its resources. This separation also divorces us from the truth of our own mortality. In Heidegger, it is one's own death that is paramount. Ignorant of our own death, we move through the life-cycle believing that we have control over all the things at our disposal; we reduce them to use-value rather than seeing their vital nature and essential contingency. The arrogance of mastery produces a disregard for the insubstantiality of the earth as well as oneself and others.

Human-induced climate change is an obvious example of this plight and its ruinous consequences: we have denied and deluded ourselves about the finite resources of the earth. We continue to insist upon dualities of nature/culture, self/other that only prolong our existential anxiety. In the face of repeated scientific reports and the global increase in severe and unpredictable weather events, we have steadfastly held onto the false certainty that the earth is permanent and will provide unconditional hospitality to the human species for evermore; that we will never die. We have subsequently thwarted our duty of care, our responsibilities of stewardship, towards the environment.

At first glance, the link between death and ecological destruction may seem abstract. Yet it makes sense that the two are connected: if we cannot confront and accept our own vulnerability and finitude, how can we appreciate these qualities in other things? Heidegger explains the relationship between mortality, dwelling and care in *Building Dwelling Thinking*, arguing that it is only because we are mortal that we are able to dwell on this earth (1978a). Tracing the linguistic roots of various Germanic terms, he argues for the enmeshment of being human with dwelling and, what is more, the link this has with cherishing and protecting the world when we 'build'; building which 'only takes care – it tends the growth that ripens into its fruits of its own accord' (1978a: 325). We find ourselves thrown into a world where we are both mortal and charged with caring for the environment. It is only through accepting these two conditions that we can accomplish a 'good death' and let go of the 'harassed unrest' that marks the plight of dwelling (1978a: 328).

DWELLING-WITH

Caring for and protecting our dwelling place opens the possibility, at a personal level, for a 'good death'. Does this care for my dwelling extend to my concerns for another's? Or more drastically, my need to welcome the homeless neighbour? Where does the foreigner, the neighbour, the stranger, the exile and migrant, sit in this formulation? In Heidegger, the 'I' is not isolated but does retain a sense of being discrete. In his earlier work, *Being and Time*, Heidegger acknowledged that Dasein, 'I', while discrete, is always with others who share a common world. Importantly, Dasein means 'being-there'. Moreover, Heidegger says that we should appreciate the 'with', of 'being-with', as an existential rather than a categorical framing, pointing to the foundational place of Others in how 'I' experience 'being-in-the-world'. Moreover, he emphasises 'circumspect' concern, pointing to a reluctance to take risks; a caution that may speak to an unwillingness to contribute to the destruction or spoiling of the world that I 'share with others' (2008: 154–5). To extend care that acknowledges finitude, and the fact of sharing the world with others, amounts to dwelling-with or adapting-well. An attitude of care should orient our relationship to place and others.

DWELLING AND THE OTHER

Dwelling-with, we recognise our own finitude as well as that of the others (their separate being-in-the-world). We care for the environment and others because they constitute our world and provoke in us a capacity for responsiveness. Heidegger helps us to create a perspective of dwelling that is more attuned to the capacity and duty of caring for oneself and one's environment, in turn respecting the dwelling of others. However, his concepts of Being and dwelling are less able to provide a strong platform for really appreciating our responsibility towards the Other as the human stranger who makes demands upon us. Heidegger tends to see the Other in terms of another Dasein ('Dasein-with' or *Mitsein*) (Heidegger 2008: 155). Heidegger's phi-

An Ethos of Dwelling

losophy emphasises the ontological structure of Being without specific articulations of ethics (Poleshchuk 2010: 7; Drabinski 2012). Care is not an ethic of the Other but an unfolding permitted when one is dwelling authentically. Care is a disposition of 'self-concern' and proper regard for the projects unfolding in one's world (Groves 2014: 101). Thus, while the notion of care Heidegger presents is useful for reorienting our personal relationship to the world and our own death, it does not particularly accommodate (literally) the foreigner in 'our' world.

A more radical responsibility towards the Other emerges out of Levinas's work on ethics and mortality. Responsibility (or response-ability) is core to Levinas's articulation of an ethics of the Other. In the face of the Other, the foreigner, the neighbour, the refugee, we are compelled to recognise our infinite responsibility for their welfare. The mortality of the stranger, rather than our own, summons us to respond: 'In its expression, in its mortality, the face before me summons me, calls for me, as if the invisible death that must be faced by the Other . . . were my business' (Hand 1989: 83). Jewish theologian Martin Kavka summarises the meaning and implications of Levinas's notion of response-ability well:

> To live life in accordance with this fact – to live in accordance with the law of what it is to be human – is to acknowledge that a life of egoism, or a life lived in accordance with a set of abstract and universal norms . . . is a life of false consciousness. Instead, one works on behalf of others to promote their flourishing and to produce justice. Thus from an analysis of our language-use as engaging with an ability to respond, Levinas deduced an ethics of responsibility. (Kavka 2013: 289)

This responsibility is simultaneous with the affirmation of the right to be. Once I say 'I', I am always already responsible for the other: 'language is born in responsibility' (Levinas in Hand 1989: 82). Language is intersubjective and indebted to the other. 'I' am indebted to the Other.

Proximity is no barrier to engaging with our infinite responsibility: it does not matter how close or far away the person in need is. Further, drawing on Levinas, it matters not if we have benefited from unexpected harms, as per ethical justifications for analytic models such as distributive responsibility. The impetus for action is in the interpersonal nature of our existence and the great debt we owe to the Other for our very being. Our relationality makes the destruction of the stranger 'my business'. In the context of climate change, this encompasses both the potential loss of life and lands, homes and homelands, as well as the agony of cultural loss and emotional suffering the victim endures. In the face of their suffering we are called to respond: suffering 'is at once what disturbs order and this disturbance itself' (Levinas 1998: 78).

A Levinasian concept of 'dwelling' is also more radical than Heidegger's proposal.[5] While we can find ourselves at-home, that is, authentically dwelling, in Heidegger, in Levinas this condition is immediately contested by the encounter with the Other (Levinas 1969). At its heart is the injunction to welcome the Other and, through this, release the egoist grip over a possessive concept of 'dwelling': 'The possibility for the home to open to the Other is as essential to the essence of the home as closed doors and windows' (Levinas 1969: 173). Levinas's philosophy invites us to awaken to the interdependency we have with the stranger. We experience the most profoundly human qualities when we give to and welcome another. Again, this is not a choice but a condition of life itself, a condition that we continually negate, erecting borders and walls, violently refusing the Other:

> It is in generosity that the world possessed by me – the world open to enjoyment – is apperceived from a point of view independent of the egoist position. The 'objective' is not simply the object of an impassive contemplation. Or rather impassive contemplation is defined by gift, by the abolition of inalienable property. The presence of the Other is equivalent to this calling into question of my joyous possession of the world. (Levinas 1969: 75–6)

The gift here is in loosening the strictures of possession. Building gives way to giving (Levinas 1969: 77). We can only perceive the potential of true generosity when we move away from, or become independent of, our 'egoist position'. The ego grows through the act of possession, yet at the same time the control that possession promises is an illusion. Thus the egoist position – like the protectionist nationalist position – is wreaked by its paranoia and anxiety. This is why it is important to recognise that part and parcel of relinquishing the grip of the ego is the radical call to abolish 'inalienable property'.

Thus we have two philosophical notions of dwelling at our disposal here. The first, Heideggerian dwelling, emphasises facing mortality in

An Ethos of Dwelling

order to centre 'care'. Only by dwelling authentically with the facticity of finitude can we embrace a relationship to the world that is caring and be at-home in our temporary dwelling. The second, a Levinasian formulation, calls upon us to disrupt any sense of enclosure the home may offer by recognising the potential death of the Other as our business; a matter that concerns me. It is productive to introduce Derrida's work at this juncture as he relentlessly straddles these theoretical positions, offering a deconstructive lens that holds place and placelessness together. It is the question of 'where' one belongs that brings these concerns together and demonstrates the impossibility of settling on either side of the dichotomy. The question 'where?' is 'ageless, transitive, it gives as essential the relation to place, to dwelling, to placelessness' (Dufourmantelle and Derrida 2000: 54). This question repeats itself endlessly because it can never rest. It is ageless in that it is not historical but metaphysical, or *a priori*: it does not matter what political or social structures emerge; as human beings we are always grappling with our mortality and thus our inevitable loss of place on this earth. Finally, it highlights the tension of dwelling itself: dwelling moves between a relationship to place *and* placelessness. It is through Levinas that Derrida's meaning at this point can be better understood. Dwelling is never just the assurance of our own ownership over a place, or orientation within. It is fundamentally about the way that one's location is always contested even before it is socially constituted. Because of this condition of existence, strangers 'disturb' the illusion that I am enclosed within my dwelling (Levinas 1969: 39). This is not a simple intellectual exercise. Levinas knows better than most what happens when people close themselves off from the Other, having survived the horrors of the Holocaust himself.

To limit ourselves to ethical paradigms that calculate our responsibilities, while assuming the separation of the self from the Other, is to restrict the full potential of our ethical creativity. Levinas's contention that the self 'cannot disentangle [itself] from society with the other' (1969: 47), goes beyond the simple fact of co-existence. Levinas regards the 'self' as borne out of the intersubjective relation with the Other. This is a relational ethics. More than this, the 'self' owes its existence to the Other. This is why the Other concerns me so profoundly. It is not a rational choice to declare my concern and divvy up the obligations. Generosity towards the Other is part of the very fibre of my being at the same time as it undoes the egoist preoccupations with possession and control (1969: 75). When this is accepted and enacted, 'ipseity [the self] is graceful, lightened of its egoist unwieldiness' (1969: 301).

AN INVIOLABLE HOME: COMMUNAL RELOCATION, THE INUIT PETITION AND THE 'RIGHT TO A HOME'

It is undeniable that adaptation presents us with many ethically vexed scenarios, especially when we consider the myriad issues it gives rise to in relation to human mobility. Within a normative lens we may ask: where should a country's obligations to its neighbours begin? Should it be at the moment when a migrant crosses a border, or beforehand, when people find themselves internally displaced due to the climate? Within a continental philosophical tradition, these questions are pragmatically necessary, but tend to bypass the foundational assumptions that generate ethical decisions. Instead, we may ask, what does it mean to 'answer for one's dwelling place, for one's identity'? (Dufourmantelle and Derrida 2000: 149–51). What does it mean to be a 'self'? Is hospitality or responsibility calculable? How do we dwell-well, or dwell-with, ethically? Indeed, these ways of framing the issue helps us to understand the significance of the divergent concerns of normative and continental thinking. Perhaps an even greater difficulty presents itself when we try to articulate forms of preventative as well as restorative responsibility in instances where communal relocation is a real possibility. In this arena, contemporary normative climate ethics, with its adherence to redistributed responsibility for harms, finds itself at a loss. Harm to-come, future harm, harm that has yet to fully arrive, cannot be accommodated within the logics of cost–benefit analysis. Yet communal relocation is one such intergenerational concern that calls upon us to activate our ethical imagination now, in the name of the future.

An example of communal relocations that have been unfolding for over a decade is the Alaskan region. In December 2005, the NGO Inuit Circumpolar Conference (hereinafter, the Inuit), lodged a petition with the Inter-American Court of Human Rights. The Inuit represents 150,000 aboriginal peoples spread across the Arctic including Alaska, Canada, Greenland and Russia. The Inuit put forward the argument that the United States should be held legally responsible for the human rights violations that many aboriginal peoples in the Arctic region currently face as a result of climate change. In particular, it was

An Ethos of Dwelling

noted that basic rights to life, residence and movement were at stake, rights to health and wellbeing were challenged by changing climates and, perhaps most interestingly, melting ice and increased permafrost in the region threatened the right to the 'inviolability of the home' (Aminzadeh 2007: 238–9). A year later, in 2006, the Court rejected the petition, declining to entertain the case altogether, citing a lack of evidence for the claims listed.

We saw in Chapter 1 that an adequate home is a right that enables the flourishing of social textures, economic livelihood and even the capacity for effective political participation. Article IX of the *American Declaration of the Rights and Duties of Man* 1948 announces that 'every person has the right to the inviolability of his home'. Inviolability refers to the indestructibility of something; its position outside the reaches of external conditions as well as its protection against interference or assault. Legal scholar Linda McClain clarifies two dominant meanings for us. Firstly, in law, inviolability refers not to the physical indestructability of the home but to the construction of the home as 'free from arbitrary intrusion by government or others'. Secondly, and distinct from the legal realm, is the cultural notion that the home is one's 'castle' or 'sanctuary', a place to which to retire from the world (McClain 1995: 203).

The loss of one's home can be accompanied by the loss of a social texture and even exclusion from the category of humanity, as well as profound dislocation from one's spiritual and cultural place of belonging. Levinas refers to an 'essential interiority', the import of the home one possesses as the place of 'hospitality for its proprietor' (1969: 157). This primary welcome is a condition of 'recollection', or the self at home: 'to dwell . . . is a recollection, a coming to oneself, a retreat home with oneself as in a land of refuge, which answers to a hospitality, an expectancy, a human welcome' (1969: 156). To have this taken away is to be rendered homeless, without protection, without the gentleness and intimacy of the home that Levinas describes as bound up in venturing outwards.[6]

Shelia Watt-Cloutier's plea, introduced at the start of this chapter, that we 'think for a moment', returns us to the significance of the interrelationship between the physical world of land and resources, and the intangible yet equally relevant cultural, social and emotional ties peoples have with place. The two go together, and united, constitute the remedy to the diagnosed plight of dwelling. Recall in Heidegger that thinking, dwelling and Being are inseparable matters. Contrary to this, in the Global North we are rarely told to 'think for a moment' about our own spiritual, emotional and cultural (dis)connection to or from the earth (that is, our plight), let alone reflect upon the relationship other cultures have as inherently valuable and worthy of preservation. Our language has taken on the dominant modality of commodification (and technologisation) of the earth, a cost and benefit economics that is unable to properly value the spiritual, emotional and cultural.

We need only look at the political decisions that are taken in the name of climate mitigation and adaptation to see that Heidegger's fundamental point is still valid today. Robin Bronen and F. Stuart Chapin outline the bureaucratic obstacles that make effective community-driven relocation in Alaska difficult to achieve in an efficient or timely manner. They tell us that United States federal funding for mitigation is assigned on a competitive scale, 'based on cost–benefit ratios' (2013: 9321). Due to the remote location and low population of native Alaskan communities, the costs are frequently high while the 'benefits' are measured in ways that disadvantage the community when applying, or more accurately, competing, for funding (2013: 9321). Benefits may include the economic viability of the community. As Maldonado et al. write:

> Attempting to explain the harm caused to individuals and communities by claiming the greater benefit to all, cost–benefit analysis is entirely insufficient because it does not include the distribution of costs and benefits and completely ignores important social and cultural factors, instead only considering economic impacts. (Maldonado et al. 2013: 606)

An instrumentalist approach to adaptation reiterates the separation between self/other, nature/culture and human/non-human, dualisms characteristic of the plight of dwelling. When Watt-Cloutier urges us to recognise the threat climate change poses to her community's distinct culture, we can hear in this the yearning for a stance that is able to open onto the future and overcome the dangerous plight that threatens her people.

In drawing attention to these aspects of climate discourse, my intention is not to efface the resilience of marginalised communities to rebuild and flourish in the face of the challenges they are facing.

Even as disadvantaged peoples assert their capacity for adaptation, it is important to register the uneven impact of the enormous injustices that condition our relationship to the earth and its inhabitants. We may all have to endure loss, but some losses are accelerated and plainly disproportionate. These injustices call 'us', all of us, but particularly those of us with the privileges of security of place (however contingent this may be), into question and demand that we respond. Responding is our 'response-ability' (Kavka 2013: 289).

What does this responsibility look like? The common world that I share with others is my concern; and part of this is the acceptance that others have the right to a home (Porteous and Smith 2001). In Newtok, the Yup'ik, an Inuit community of 400 people, comprising 60 houses, have been preparing for communal relocation since the early 2000s (Maldonado et al. 2013: 607). Newtok is on the West Coast of Alaska, located on low-lying mud flats of the Yukon-Kuskokwim Delta and surrounded by the Ninglick River and the Bering Sea. It is expected that by as early as 2017 the community will be lost to erosion; a process sped up by anthropogenic climate change. Indeed, at the time of writing in mid-2017, the people of Newtok had lodged an application for disaster relief with the US Department of Homeland Security, requesting funding to cover the relocation of the entire community (Waldholz 2017).

Fifteen kilometres away and across the river on higher ground, a region has been confirmed as the site for the community to move to. The community have provided the name Mertarvik (meaning 'getting water from the spring'). The expected cost of this move is $130 million. Only $12 million had been received between 2009 and 2013 (Goldenberg 2013). Funding has been tied up in various bureaucratic and legal knots, while maintenance of the existing infrastructure at Newtok has stalled. As a result, the community finds itself in a state of being in-between places without any assurance of a safe, productive or sustainable future. Remaining in Newtok is an impossible scenario. Goldenberg (2013) conveys the psychological effects of this: 'Anxious residents want to know that their future is safe. They are exhausted by the years of uncertainty and fed up with a village left to decay, with leaders' energy and every scrap of funding focussed on the relocation'. Over 11,000 km away in my home in Sydney, Australia, it is easy to turn a blind eye to this situation and deny my contribution to the conditions that gave rise to the impending dispossession.

Normative ethical reasoning distinguishes between causal and moral responsibility. If a judgement is made that someone is causally responsible, it is said that they are subject to fault-based liability: guilt is identifiable and punishment for the harms inflicted (or predicted to arise) follows from this. Moral responsibility carries with it what is called strict liability; a sort of promise to make amends for harms done, despite the absence of causal responsibility (Baer 2010). Drawing on the logic of causal responsibility I could throw my hands up and declare that it is virtually impossible to draw clear lines of cause and effect between the erosion that this particular community is facing and the carbon emissions I individually am, and Australia as a state (in my case) is, responsible for now, or that Australia has been since colonisation in 1788. Moral responsibility may get us a little closer. Using strict liability criteria we may be able to collectively compel the government to contribute economically to adaptation funds as well as lobbying for formal recognition of the costs and needs of mobility issues associated with climate change (in this case, relocation). This would align well with the framework of 'common but differentiated responsibility' and offer, hopefully, some relief for disadvantaged peoples. We could reasonably conclude that we must answer 'yes' to Paul Baer's question: 'if there are unexpected harms from some activities, shouldn't the party that has benefited from the actions bear the costs of the harm, rather than the victims?' (2010: 250).

What Schroeder aptly describes as the distinct 'tone' of continental theory is recognisable if we respond to the ethical dilemmas communal relocation produces using Heidegger and Levinas. Taking a Heideggerian approach, we would need to first acknowledge, deep within our very being, that we are contributing to the spoiling of another part of the world and alongside this, we are ignorant of our constitutive condition of being-with others. Until we do this, we will see no reason to call upon our government to contribute more and support mobility options, including a communal relocation 11,000 km away. Instead, we will remain caught in a more constrained model of individualism and selfhood which is radically disconnected from the Other; a self-interested individual who has severed the bonds contained by the 'with' of being-with.

Utilising Levinas's ideas, we find ourselves in a difficult place. On the one hand, there is a need to argue for the absolute right to a home, an inviolable home for the Inuit peoples, yet on the other hand, Levinas's

An Ethos of Dwelling

thesis convincingly undoes the very possibility of any irrefutable dwelling place, as such. How can I use a theory that argues against the possibility of dwelling as uninterrupted dwelling while also calling for the recognition of the Newtok community's right to a home? Are these theoretically and practically incompatible? The short answer is 'no'. This is precisely where the deconstructive movement between place and placelessness comes into play. In Chapter 1, I introduced Derrida's neologism 'ontopology' to unpack the conflation of identity and place. I argued that current adaptation frameworks have an implicit assumption that the 'local' habitat is the most desirable location to adapt to. Moreover, in some instances, this call to remain-in-place is the most responsible position that can be taken (and this is particularly the case when we review Indigenous calls for climate justice given the imminent possibility of relocation for many across the world). In other words, reiterating the ontopological may be the best option we have available to us politically and ethically, despite the potential dangers it signifies (specifically a reliance on nativity).

This quandary has been analysed elsewhere, with 'negotiation' posited as the way forward: we cannot abandon the home (or the at-home), no; we must negotiate it (see Bulley 2006). This is a Derridean perspective but finds its inspiration in Levinas. While Levinas emphasises the transcendent, this is, as Drabinski notes, a paradoxical formulation: 'Levinas articulates transcendence as radically transcendent, wholly other, and yet, retains a sense of relationality' (2012: 84). I will cite Levinas at length here as what he writes informs the deconstructive analyses to come in the remainder of the book:

> But the transcendence of the face is not enacted outside of the world, as though the economy by which separation is produced remained beneath a sort of beatific contemplation of the Other (which would thereby turn into the idolatry that brews in all contemplation). The 'vision' of the face as face is a certain model of sojourning in a home, or – to speak in a less singular fashion – a certain form of economic life. No human or interhuman relationship can be enacted outside of economy: no face can be approached with empty hands and closed home. Recollection in a home open to the Other – hospitality – is the concrete and initial fact of human recollection and separation; it coincides with the Desire for the Other absolutely transcendent. The chosen home is

the very opposite of a root. It indicates a disengagement, a wandering which has made it possible, which is not a *less* with respect to installation, but the surplus of the relationship with the Other, metaphysics. (Levinas 1969: 172)

This passage brings to the fore the necessity of negotiation as the movement between the at-home and the contestation of my homeliness. This negotiation is informed by the primary ethical impulse that displaces the ontopological conflation of identity and place. We cannot, metaphysically, take root on the earth even as we locate ourselves in a 'chosen home'. My dwelling is always in relation to the Other who concerns me and to whom I must respond: my 'wandering' which results from the 'surplus of the relationship with the Other'. The Inuit Petition, as well as the case of Newtok, does not ask that I give up my home, that is, welcome the community into my place. Rather, it brings to light the primacy of the welcome that one receives when returning to one's private abode and the radical contingency of this; its exposure to the elements. The call for justice, in this case, is the desire to have a secure home available to the community members. It is to install as a matter of priority the hospitality one (ideally) receives when entering one's own home, what Levinas refers to as the refuge from the world that we are provided in the home (1969: 157). A deconstruction of dwelling holds together the double imperative that we simultaneously find a way to secure the homes of the Inuit community while calling our own possession into question. This deconstructive spirit is found in the above quotation from Levinas. When Levinas refers to the face-to-face relation as a 'certain model of sojourning in a home' he is bringing to attention, from the outset, the movement towards the Other that is implicit in dwelling. Moreover, he is pointing to the import of having a place that extends refuge or hospitality to its occupant; part and parcel of the 'economic life', a life that we cannot simply throw off.

CONCLUSION: DECONSTRUCTED GROUNDS

In the manner that Heidegger explicates, dwelling is about how we live, what we 'think' constitutes the basis of a worthy human life (glossed by Heidegger as 'authenticity'), and how we 'build' for the future. In order to be authentic, all of this must take place with a clear understanding of our personal, communal and planetary vulnerability and mortality. The

An Ethos of Dwelling

qualifier 'human-induced' clinches the importance of our responsibility to take heed of this and act accordingly. Yet our contemporary analytic framing of the issue means that 'responsibility' is so disputed a concept that we find ourselves stalling, squabbling over what degree of fault we can be charged with. Responsibility has been reduced to calculation.

I have elaborated in this chapter the consequences of this: the refusal to engage with the overriding concern for the 'Other' that the call to be responsible actually invokes. By illuminating the relevance of the conceptual logic of dwelling, I reinsert the imperative to care for the Other (both human and non-human) as the absolute, irrefutable foundation of our ethical responsiveness. What, then, should I do? The challenge of adapting to climate change is accompanied by an opportunity to contemplate how we dwell with others, individually, communally and nationally. It is only partially true to posit that adaptation (and dwelling) raises vexed ethical dilemmas regarding the distribution of responsibility. Drawing on a continental ethics, the situation is much more critical: the imperative to adapt contains within it a fundamental ethical injunction. First and foremost, the need to adapt actually calls our dwelling into question, and in so doing summons us to respond to the needs of the Other. We must answer for our dwelling place. In order to act with care towards others and the environment, we must critically reflect on our own stakes in possession and control. If we can do this, we may be able to recognise the common, yet unevenly experienced, vulnerability that we all share as mortal beings. In political contexts that rely on inciting hatred and fear in the community, this type of personal transformation is itself a political and ethical act. It is only by adopting this ethos that truly transformative social action can take place.

In the next chapter, I outline what it might look like if we applied normative CBDR logic to the issue of climate migration. This framework has the potential to encourage nation-states to devise climate migration programmes and special migration schemes, and even embrace the possibility of extending some sort of protection framework when responding to forced climate migrants spilling across borders. But what it keeps firmly in place is the discretion of the nation-state. Thus, I turn again to the continental tradition, building from the grounds I have developed here to argue that the foreigner question 'where?' undoes us from the inside out, and calls upon us to place the migrant at the centre of our political and ethical actions. As we will see, the ethical relation that Levinas outlines is always already complicated by the fact of there being more than one Other who disturbs my self-possession. It is not only the Inuit peoples who are turning to the global community seeking a home. Rather, what Levinas calls the 'Third' (there is always more than one another), is present from the beginning: 'the third party looks at me in the eyes of the Other – language is justice . . . the epiphany of the face qua face opens humanity' (1969: 213; see also Ziarek 2001: 66). How can we meet the challenge of ethics in a world where many Others are seeking justice?

Chapter 3

THE MOBILITY AGENDA

'The rights and the dignity of millions of fellow human beings will be further diminished if they languish in camps or on the margins of cities without access to basic needs, livelihoods and income opportunities.'

Ban Ki Moon[1]

In Chapter 1, I sought to anchor our discussion by outlining the political construction of adaptation. I examined the commitment to the 'local' in adaptation policies and in turn highlighted the implications this has for questions of identity and place. I have demonstrated that there are multiple reasons for starting with *in situ* adaptation, some economic and political, others cultural, social and ethical. This includes the cultural and emotional pull of 'roots' for some peoples; hence the Kiribati position that migration is a 'last resort'. In Bangladesh too, some have argued that 'a close sense of attachment to land, family and culture inhibits movement abroad' (McAdam and Saul 2010: 244). In this way, it makes ethical sense to support in-country adaptation. A further factor is the recognition that a continuous place of belonging, alongside economic opportunities, can provide social and psychological stability. This explains the role that the development sector plays in facilitating the implementation of adaptation projects aimed at helping people stay put, or 'safe and in place' (Martin et al. 2013; Walsham 2010).

However, by bringing to light the geopolitics of the Pacific as an example, I have suggested that adaptation policy cannot by-pass the cultural and ethical aspects of place, dwelling and mobility. Moreover, I have alluded to the possibility of being in one's 'place' and simultaneously being unsafe or disturbed by what the future holds (as in the case of the people of Newtok, Alaska). I have argued that our articulation of adaptation must grapple with the complex nature of human mobility. Examples of mobility range from chronic internal displacement, immobility or experiences of being trapped in a place, forced and 'voluntary' migration, and communal relocation. These issues will not simply disappear and must, at least to start with, be considered under the adaptation umbrella. The fact of widespread and growing human movement forces us to confront the relationship between place/-lessness, identity and ethics anew. In Chapter 2, I opened this line of inquiry, developing some of the philosophical resources continental theory has available to us for such an essential endeavour. Inspired by Levinas and Derrida, we must give attention not just to the condition of homelessness the foreigner finds him- or herself in, but also to the manner in which we constitute our 'home' through our defence of arbitrary boundaries. We believe that these lines over the earth solidify the entitlement 'we' have to 'our' place. This belief has significant consequences.

It is now necessary to consider in greater depth the political and ethical issues that accompany the emergence of a climate-mobility agenda and its relationship to adaptation discourse. In recent years, international dialogue concerning adaptation has expanded to include mobility (Piguet et al. 2011; Faist and Schade 2013). Alongside this, however, it is impossible to ignore the strong undercurrents of anti-immigration and xenophobia that are spreading across the globe, particularly following the Syrian crisis in Europe in 2015, the Brexit vote in the United Kingdom, and the election of President Donald Trump in the United States, both in 2016. Immigration issues especially give rise to a struggle over social meaning and values. This struggle can be summarised by two dominant positions. Firstly, there is a security discourse which promotes the need to regulate human mobility and protect national borders (protectionist politics). Secondly, human-centred approaches advocate for the human rights and human security of the many millions of people displaced, dispossessed or migrating.[2] Just as this duality marks mainstream migration attitudes, it also characterises the issue of climate mobility. Added to this are enormous gaps in legal infrastructure which continue to exist with no enforceable international protection protocols or guidelines related to climate-driven migration. As this chapter unfolds, I will develop an argument in favour of both broad mobility rights as well as a protection regime for forced climate migrants.

As I will demonstrate, the national security or human security duality is sustained by a notion of politics that starts with the nation-state and the citizen (Nail 2015). Politics contributes to how we relate to place and provides the paradigms we use to understand human movement, such as citizen/foreigner, authorised/unauthorised, legal/illegal and regular/irregular (Soguk 1999). This has implications for how we practise ethics too: when we start with the citizen and the state, our ethical paradigms attempt to convince state institutions to act out of obligation. It is this logic that underpins attempts to encourage member states to sign up to the ethical paradigm of CBDR. However, because of the assumption of stasis that the political community of the state desperately seeks to reproduce, ethical action meets crippling obstacles. States would much prefer it if we just stay put or move only with explicit, pre-arranged permission! Such assumptions of political theory have been usefully critiqued by Thomas Nail who urges us to consider reformulating politics in a way that takes the migrant and mobility as the starting point, rather than the citizen and the state. This is a radical invitation and one that prompts us, I believe, to reconsider and re-energise our ethical vocabulary as well.[3]

MIGRATION AS A MEGA-TREND

The worthy ambition to assist people to stay 'safe and in place' is no longer possible for many people who find themselves on the move. This can no longer be ignored or relegated to the worries of the political left. We must revitalise our ethical and political discourse in a way that centres on the migrant. So significant is the role of climate mobility that the IOM Director General William Lacy Swing has referred to migration as 'a "mega-trend" of our century' that will continue into the future (IOM 2015c). This movement is complex in relation to timeframes (short or long term), trigger points, and causality (economic, political, climate, environment) and, most significantly, destinations, with most migration internal, some regional, and a smaller percentage cross-border (McAdam 2011; Laczko and Aghazarm 2009; Walsham 2010).

In the Asia-Pacific, for instance, numerous regions are designated climate change 'hot spots'. These are areas where local adaptation will very likely prove an inadequate strategy, risking the possibility of mass internal movements from rural to urban regions, as well as, to lesser degrees, regional and international migrations (Walsham 2010: 6). In South-East Asia, frequent and increasingly severe flooding, rising sea levels, and other hazards such as earthquakes, droughts, typhoons, landslides and tsunamis affect the region. For example, in 2014, the Philippines suffered enormous levels of displacement following Typhoons Rammasun and Hagupit, with almost five million people displaced (Nansen Initiative 2015: 25). Regional cross-border migration is common: in 2010, Thailand was the recipient of Cambodian migrants seeking a livelihood following prolonged drought and food insecurity, while in 2011, Thailand was also the destination country for over 100,000 Burmese fleeing the impacts of extreme flooding (2015: 27).

There is little argument that we are highly mobile societies, indeed, we are an intrinsically mobile species. In a world of seven billion people, approximately one billion are classified as migrants of some sort. Globally it is estimated that close to 60 million people are refugees, asylum seekers or internally displaced peoples (IOM 2015a). Predictions point to a further 200 million people displaced, many internally, due to climate change in the coming decades (IOM 2015b). This is not difficult to imagine when the International Displacement Monitoring Centre says that between 2008 and 2014, close to 185 million people were displaced due to disasters (climate related and other) (Nansen Initiative 2015). The IPCC 'Working Group 2 on Adaptation' has noted that climate-related mobility may very well 'become a defining humanitarian and development issue in the coming decades' (Warner et al. 2014: 11). Given this, rather than viewing movement (in its myriad forms) as the aberration, we need instead to view it as a historical constant (Arendt 1951; Nail 2015).

Despite knowledge of the prevalence of movement globally ('organised' and 'unorganised'), all forms of mobility are sensitive and intensely politicised issues. In relation to South to North migration movements, this politicisation tends to take the form of securitisation. Securitisation refers to a 'social practice' which involves constructing an issue as a threat (Trombetta 2008: 588; Boas 2015). In relation to climate change this is evident in the fear-driven rhetoric of waves of 'environmental refugees' and 'climate refugees' which has been part of the discourse since the prominent environmentalist Norman Myers unhelpfully characterised environmental migration as pathological and destructive (1993: 200–1). This anti-immigration sentiment is evident in debates across the Global North.[4] Dressing old racisms and xenophobia up in fear-laden discourses of environmentalist protectionism is on the rise (Gourevitch 2010: 412).

Populist positions like that of Myers hinder the communication of work which attempts to navigate and explain the complex scenario we actually face. Climate change takes place in social and political environments with pre-existing inequalities as well as patterns of migration as multifaceted as the contexts themselves. In Africa and the Middle East, climate variation and extreme weather such as droughts and flooding exacerbate regional conflicts and add pressure to already unstable governance structures. For instance, in Central Africa, severe weather has worked in tandem with conflict and insecurity, wearing down local resilience and pushing an unknown but significant number of people across regional, rather than international, borders (Nansen Initiative 2015: 9–10). During 2011–12, chronic drought conditions in Somalia contributed to the displacement of almost 300,000 people internally, while another 300,000 moved across borders into the neighbouring countries of Ethiopia and Kenya (2015: 10).

Worldwide, the greater part of movement is regional and falls upon the least advantaged societies to shoulder. In fact, the global refugee issue highlights the stark inequities that continue to inform international 'burden sharing' of refugee populations.[5] In 2015, the United Nations High Commissioner for Refugees (UNHCR) issued a report explaining that:

> The developing regions continue to receive refugees disproportionately, with most hosted by low- and middle-income countries. For three years in a row, countries in these regions have hosted an average of 86 percent of all refugees under UNHCR's mandate ... the Least Developed Countries – those least able to meet the development needs of their own citizens, let alone the humanitarian needs often associated with refugee crises – provided asylum to over 4 million people. (UNHCR 2015)

As long as the countries of the Global North continue to perceive mobility as a threat, rather than an ordinary human practice, they will be tempted to ignore the empirical realities and take refuge, instead, in the false promise of protection offered by a security response. Viewing migration as a 'threat' reiterates the notion of migration as an aberration of the sedentary or settled status quo.

The securitisation of migration is being challenged by a shift in the mainstream adaptation discourse to include migration. On the one hand, securitisation is bound up with asserting national interests and fostering an anxious citizenry lacking any ethical concern for vulnerable peoples on the move. On the other hand, migration-as-adaptation is informed by rights, human security and broader ethical principles of intergenerational responsibility and differentiated responsibility. This platform attempts to normalise and regulate diverse forms of mobility. In other words, an international climate mobility agenda is emerging, informed by normative ethical ideals, and particularly distributed responsibility and human rights, which seek to work with, as well as challenge, the organising principle of security.

THE MOBILITY AGENDA

If we take a long view of migration, outside the reaches of the strictly 'historical' and into the geological record, we see that human communities have regularly moved in response to changing climate conditions. Deep history approaches reveal that modern humans (*Homo sapiens*) may very well have migrated away from Africa some 100,000 years ago as a direct response to climate change (Carto et al. 2009; Burroughs 2005: 100–15; Manning 2005). Along the Eastern and Southern coastlines, communities likely moved in response to changing availability of resources or environmental conditions. Even before this, *Homo erectus* had needed to respond to changing climates using innovative techniques in order to survive; they 'had to learn to detach experiences from particular places in their familiar environment and to find new, comparable locations' (Behringer 2010: 30).

The Neolithic era of human settlement emerged in the early days of the Holocene, the geological period we are currently in and one marked by its relative climate stability. For Behringer, the Neolithic 'represents a decisive period in the history of humanity: the passage from the semi-nomadic hunter-gatherer culture of the Middle Stone Age to a sedentary culture of farmers and livestock breeders' (2010: 45). However, rather than viewing sedentary culture as static, it is helpful to recall Nail's reframing of social movement, which is able to take account of the historically complex nature in which moving populations came to be simultaneously settled, and permitted to circulate within proscribed limits (the village, city-state, national territories and so forth) and for specific purposes (labour, for example) (2015: 39). Thus, paradoxically, the development of a politics of social movement, as Nail argues, 'emerged with the first sedentary human societies' (2015: 39).

A geological lens confirms the explanatory power of 'migration as adaptation'. Of course, it is not sufficient to offer up a scientific perspective of migration as characteristic of human adaptive responses to the climate. The fact of adaptive migrations extending back 100,000 years does not change the historical specificity of contemporary migratory flows and the political responses they provoke. In this way, within contemporary discourse 'migration as adaptation' is a distinctly political construction. As a historical phenomenon, 'migration as adaptation' arose out of debates in the 1980s concerning 'environmental refugees' (El-Hinnawi 1985). In the 1980s, fear of mass South–North migrations prompted calls for action on environmental issues to prevent the flow of homeless peoples across borders. Within academic and policy debates 'causality' quickly emerged as a problem. Legal classification of movement requires conceptual demarcations based on clear lines of cause and effect. Over the next two decades, empirical research convincingly demonstrated that the environment and/or climate is rarely the sole driver of migration, but is nonetheless part of the story (Piguet et al. 2011). As the evidence settled, the work of providing frameworks of protection, rights and security became entangled in ongoing disagreement over causality and, from there, problems related to the distribution of responsibility. Thirty years later and the case of any link between the environment and migration 'remains a black box' issue (Hillman et al. 2015: 2). Consequently, no agreements have been reached over categorisation or responsibility.

As scholars teased out the conceptual complexities of the subject, the IPCC had already recognised that climate-driven migration would be a global phenomenon and an unavoidable political reality with ethical consequence (Brown et al. 2007). Initially on the outer margins of discourse, international negotiations concerning climate-related mobility changed significantly over the timespan of the meetings attended by the Conference of Parties (COP). COP is the UNFCCC member states' annual conference where policy agendas and political decisions are made, the first of which was held in Berlin in 1995. Early conferences focused on mitigation measures, urging participant states to sign up to emission reduction targets in accordance with the *Kyoto Protocol* 1997 (which came into force in 2005). By the time of the seventh meeting in Morocco, 2001, mitigation could no longer guarantee the livelihood and security of many people around the world. Simply put, it was now recognised that it would be impossible to prevent the impacts of climate change from affecting human settlements, especially those categorised as 'least developed'. Moreover, it was acknowledged that, much like action on mitigation, the UNFCCC's adherence to CBDR also needed to apply to financing adaptation as well as the equitable transfer of technologies to countries in need (Dellink et al. 2009; Grasso 2010; Honkonen 2009). As such, in 2001, an LDC advisory committee was organised, with NAPs established as the best way for vulnerable countries to identify their adaptation needs and seek international funding.

The COP *Cancun Agreements* 2010 explicitly introduces the relevance of human mobility in relation to climate change and adaptation. Article 14(f) calls for 'Measures to enhance understanding, coordination and cooperation with regard to climate change induced displacement, migration and planned relocation, where appropriate, at the national, regional and international levels' (2010: 5). In response, an Advisory Group on Climate Change and Human Mobility was set up and produced recommendations to the international community, set out in its report, *Human Mobility in the Context of Climate Change* (2015). For the UNFCCC member states to support an advisory group signalled a clear intention to take climate mobility seriously. The report puts forward three primary and remarkable recommendations. Firstly, it calls upon the international community to 'strengthen the resilience of climate vulnerable populations to enable them to remain where they live'; secondly, it suggests that there needs to be support for 'voluntary and dignified internal and cross-border migration as an adaptation strategy'; and thirdly, noted as a last resort, yet nonetheless named as necessary, is planning for 'participatory and dignified relocation' (2015: 3). The qualifier 'dignified' works to remind policymakers of the importance of human rights frameworks in the development of 'facilitated migration and planned relocation' (2015: 7). The proposed initiatives work within the state-based system: they seek to reorient the state's relationship to climate mobility issues by casting the issue in terms the state deems appropriate, such as 'facilitated' and 'planned'. Moreover, the proposal ties this call for state engagement back into international rights norms: thus migration and relocation are 'participatory', 'dignified' and 'voluntary'.

Two significant issues remain unspoken in these discussions. Rights to mobility are weak and, related to this, there is the potential need to develop specific climate change protection frameworks for people whose mobility is categorised as 'unregulated' or 'illegal'. If migration is 'forced', a protection agenda should be available to assist with the safe passage of vulnerable peoples as well as a place of refuge at the other end of the migration (Biermann and Boas 2010; Ferris 2011; Nansen Initiative 2015). That is, you should have a 'right to a safe migration' when forced to move, flee, relocate and so forth. In addition, there is a growing need to develop a 'right to refuge or resettlement' for those displaced and forcibly migrating.[6] While climate change certainly blurs the dichotomy between voluntary and forced, distinctions in legal categorisation still tend to demand a hard line be drawn in order to make decisions about the distribution of resources and the uptake of applicable rights and appropriate services.

Despite these shortcomings, the COP climate change talks in Paris 2015 were anticipated to be a turning point in the international discourse on climate mobility. The current situation and forecasts warranted that the label 'crisis' be applied to human mobility concerns. In turn, this compelled the global community to take action. Geographers Christine Gibb and James Ford rightly point out that the *Cancun Agreements* identifies the UNFCCC as an 'appropriate forum for pursuing climate displacement, migration and planned relocation' (2012: 1). While the UNHCR has stretched its resource base to assist people displaced by environmental disaster in Myanmar, China, Pakistan and Bangladesh, it does not possess an institutional protection mandate beyond that identified under the *United Nations Convention Relating to the Status of Refugees* 1951 (Martin 2012: 1051). Placing responsibility under the auspices of the UNFCCC is significant because it engenders a global response to climate mobility issues informed by the ethical principle of a CBDR. In other words, mobility would be a global ethical responsibility. Moreover, and in accordance with principles of equity, Global North states would be handed a greater responsibility for redressing harms and minimising risks associated with mobility.

The draft report of the Advisory Group on Climate Change and Human Mobility proposes the development of desirable models for organising human movement in order to minimise human security risks as well as assuage the national security concerns with regards to unregulated movement and threats to sovereignty. By discursively (if not through implementation) mainstreaming migration as adaptation, as well as canvassing broader mobility challenges the global community will face, two core ethical norms are dominant: distributive justice and human rights. As such the draft report urges the international community to develop categories of, and governance models for, the authorised and managed process of adaptive climate mobility.

Within this context, mobility is no longer viewed as a failure to adapt. In place of the language of failed adaptation we come to view migration as *manageable*, much like a chronic disease.[7] Instead of seeing migration as a problem to be rooted out of the global system (acknowledged to be an impossibility), it is taken up as an issue to be thoroughly regulated and controlled. Once we can handle this condition adequately we can move away from a focus on securing borders to providing 'migration with dignity' within a system of nation-states (McAdam and Saul 2010). On the one hand, this is a very worthy move forward in the international dialogue.[8] On the other hand, management of a social issue comes with its own risks, with border control banners already used to justify a range of restrictive and exclusionary state practices throughout the world (Vaughan-Williams 2015).

Diplomatic gains made in the international arena regarding migration as adaptation have not gone uncontested. There has been strong resistance from the Global South, particularly LDCs, to placing mobility within adaptation frameworks. This seems, on the surface, counterintuitive. Should disadvantaged states not feel some relief that displacement, migration and relocation is beginning to gain acknowledgement as important internationally? Perhaps these states are rightly weary of what 'management', guided by the interests of the Global North, looks like. Thus what emerges is a split in thinking: on the one hand, the Global North, or wealthy states, wish to frame mobility as adaptive, while on the other hand LDCs argue that the issue of mobility should be placed under the radical and more contested umbrella of Loss and Damages, thus incurring substantial financial compensation and other forms of legal culpability.[9]

Against the backdrop of these debates, leading up to the Paris Conference of December 2015, Australia successfully lobbied for the removal of the *Human Mobility in the Context of Climate Change* report from the COP agenda. Australia challenged the move away from local adaptation, citing its financial pledge of $200 million to the GCF as well as the $50 million it has dedicated to the Pacific for resilience-building *in situ* (Karasapan 2015), discussed in Chapter 1 of this book. Despite the voluntary and non-binding nature of the Cancun Adaptation Framework, Article 14(f) was, we must conclude, considered by Australia to place too great a responsibility on the Global North countries for dealing with mobile populations. The reluctance to demonstrate political willingness to reasonably negotiate safe and humane solutions to mobility issues is not restricted to Australia. As United Nations Secretary-General Ban Ki Moon (2007–16) notes, even with the current rights frameworks and more than sufficient policy recommendations already available for use, across the world 'securitisation and closure of borders' continues alongside the 'growing trend of criminalisation of irregular movements' (United Nations General Assembly 2016: 14–15; see also Taran 2001).

Ethics takes a back seat to politics. There is a turn away from the question 'where?'

COMMON BUT DIFFERENTIATED RESPONSIBILITY AND THE CLIMATE MIGRANT

If we take adaptation to refer to either survival techniques or transformative practices aimed at ensuring human futures, mobility in all its forms is an adaptive action. What remains up for debate is how we respond locally, regionally and globally. Who is responsible? What ethical principles should guide our political response and indeed our 'distribution' of responsibility? Which institutions should be responsible for governance?

Historically, legally and politically, coupling mobility (barring matters of asylum) with ethics (however conceived) has been pushed away in favour of subordinating immigration matters to the authority of state sovereignty. There are no universal rights to mobility, and indeed rights to movement are guaranteed only within one's nation-state (and in some states, even internal movement is tightly controlled). The difficulty many face finding a new home demands more from us than the worldwide expansion of detention facilities. Migration has not been framed as a primarily ethical consideration, but a political, legal and economic matter for states to negotiate domestically. As a result, applying CBDR logic to the issue of climate-related mobility is both unpopular and deemed, I would suggest, too threatening to the control that states like to assume over who can enter, and how people cross, political borders. In the Australian bureaucratic lexicon, for example, this is blandly termed 'border management'; a weasel word that disguises the violence inherent to the task of determining inclusion and exclusion.

Are normative ethical principles, such as distributive responsibility, taken to breaking point in relation to the issue of global migration? No, but they do need to be challenged by alternative ethical ideas that reinforce the non-deferrable nature of our responsibility. 'There is nothing to do but find another home', laments a Bangladeshi village elder following the devastation of Cyclone Alia in 2009 (Kartiki 2011: 28). While refugee movements have prompted renewed calls from the UN for 'responsibility-sharing' between member states (United Nations General Assembly 2016), the absence of special categories of movement regarding climate change means that a protection agenda does not currently compel states to provide refuge to victims of climate disaster. Individuals and communities cannot claim a special right to protection from another state.

Consequently, efforts to apply CBDR are based on arguments that start with appeals to obligation on the part of the Global North states to welcome and accommodate the victimised stranger. CBDR responds to the question 'where?' by calculating degrees of responsibility. The calculation of responsibility follows on from assessing harm, damage and capacity. Once the degree of responsibility is determined, the receiving state must accept its duty and, so to speak, open the door to the stranger. Because causality is irresolvable and used by states to deny responsibility, normative ethical reasoning distinguishes between causal responsibility and moral responsibility (discussed in Chapter 2). The UNFCCC does not establish legal or fault-based liability but it does offer up moral responsibility in its articulation of CBDR. Philosopher Paul Baer nicely surmises the question at the heart of calls for strict liability when he asks, 'if there are unexpected harms from some activities, shouldn't the party that benefited from the actions bear the costs of the harm, rather than the victims?' (2010: 250).

Redressing harms is not necessarily the same as offering hospitality to the homeless stranger. This gives rise to another question: is the Global North accountable to the world's climate-displaced and responsible for providing a new home (not just economic compensation for harm)? Sujatha Byravan and Sudhir Chella Rajan have responded with an emphatic 'yes' to this question:

> Under our proposed framework people living in areas that are likely to be obliterated or rendered uninhabitable would be provided the early option of migrating legally in numbers that are in some proportion to the host countries' cumulative greenhouse gas emissions. (Byravan and Rajan 2006: 249)

In other words, higher historical emissions equates with a greater 'index of responsibility', especially once it is determined that the potential host country has the capacity (resources/wealth, services, space) available to assist (Baer 2010: 252). Byravan and Rajan regard the 'polluter pays principle' as transferable to the context of migration. 'Capacity' will inevitably emerge as a contentious condition. For instance, high-emitting countries such as the United Kingdom should, within this reasoning, prepare themselves to accept many thousands of climate migrants over the next few decades. Yet the Brexit decision of 2016, driven as it was by an anti-immigrant sentiment, makes this appear both difficult and unlikely. One could argue that the United Kingdom lacks the moral capacity to humanely welcome climate migrants. Imposing moral justifications from above is not going to change the grassroots-level treatment of foreigners in the community. It is for this reason that we need, also, to expound ethical paradigms that begin with the individual person and invite a personal transformation: a reflection of one's own dwelling in both the Heideggerian and Levinasian senses.

Moral responsibility for the displacement and even dispossession of people due to climate destruction has been expressed by LDC states too. In 2009, former Bangladeshi Finance Minister Saber Chowdery told the international media that what was needed was:

> a protocol that facilitates managed migration, not necessarily [for] the person who has been displaced, but let's say for every ten individuals have been displaced, you know, a country in the West, or an industrialised country, takes one immigrant. And it can be a win-win situation. Let's say that particular host country is in need of nurses, medical care. And we in Bangladesh can then train those people and send them out. (Chowdery in Carrick 2011)

This argument draws on an ethics of intergenerational and distributed responsibility as well as positing a cause and effect logic which holds the Global North causally, not just morally, responsible. A responsible agent can be located and held to account, without having to revise rights frameworks. Managed migration along skill-based lines is an attempt to create 'legal' migration patterns with social justice appeals that recognise the unequal impacts of climate change. Yet because this ethical reasoning requires abstraction in order to generate general principles, the specificity of context is ignored. Who is migrating? Not necessarily the person displaced internally in Bangladesh, as Minister Chowdery says. In Chapter 4, I take up the case of Bangladesh in detail, exploring the issue of international migration and ethics. I tease out CBDR alongside a more 'radical' ethic of (un)conditional hospitality (inspired by Derrida) and informed by a social justice platform that advocates for a 'pro-poor' economic migration agenda.

CBDR is already contentious as a principle in that it places moral pressure on the Global North states to pledge financial contributions to *in situ* adaptation funds. Yet we find ourselves bound to this ethical paradigm. We are convinced of its efficacy and pragmatic qualities. The trouble is, these normative principles, for all their validity, worth and necessity, take for granted the absolute authority of sovereign statehood. As Schroeder explains, 'applied ethics takes the basic functions of an institution for granted; it resolves tensions and hard cases by ordering those functions' (2013: 462). Can we take for granted the potential for the state to live in accordance with principles that sideline self-interest? We can, but we will also be subjected to the potential for violence that sovereignty contains within it as a structuring condition of its expression (Arendt 1951; Mansfield 2010; Hägglund 2008).

AN OTHER-ORIENTED HUMAN RIGHT TO MOBILITY

As I have been arguing, a political theory that emerges out of the primacy of movement and the migrant needs an ethical orientation that takes the question of 'where?', the condition of the migrant, dispossessed and displaced, as a priority. 'Where?', you might recall, is considered the primary question of humankind, according to Dufourmantelle and Derrida, because it compels us to embody the uncomfortable and irresolvable relationship between place, placelessness, the self and ethics. In this section and the next, I will argue that, informed by a Levinasian ethics, human rights approaches to migration and its governance begin with the needs of the migrant rather than the nation-state. Further, and building on this foundation, international protection agendas get us closer to a relational ethics of dwelling that embodies the reality of mutual vulnerability, undeniable interdependency and asymmetrical responsibility.

Indeed, we can detect the desire to prioritise the migrant in formal statements issued by the United Nations Office of the High Commissioner for Human Rights: 'A human rights-based approach to migration places the migrant at the centre of migration policies and governance, and pays particular attention to the situation of marginalised and disadvantaged groups of migrants' (2016). At present, however, the rights of people migrating, displaced, temporarily or chronically uprooted or mobile are scant. We have not achieved anything resembling this ideal outlined by the United Nations. This is not to say there are no rights to mobility codified in international covenants. If member states have signed and ratified the various doctrines available, a number of mobility rights are recognised. There are, for example, rights to the freedom of movement within the borders of one's state. Article 13(1) of the *Universal Declaration of Human Rights* 1948 names this, as well as the right to leave and return to one's country of citizenship (Article 13(2)). This amounts to a right to emigrate, without a corresponding right to immigrate, a contradiction noted by Pécoud and de Guchteneire as an ethical concern requiring attention (2006: 75; 2007). Article 14 specifically references the right to asylum, outlined comprehensively in the *United Nations Convention Relating to the Status of Refugees* 1951 (and the 1967 amendments).

Significantly, the *Refugee Convention* excludes 'non-political crimes' as constituting the basis of a claim to asylum; a qualification that has given rise to many debates, including those relating to environmental and climate-driven displacement and migration (McAdam 2011; Biermann and Boas 2010). This returns us to the split in perspectives regarding social, economic and cultural rights, violations of which do not (without a political dimension) constitute a claim for refuge. Most recently, the *International Convention on the Protection of the Rights of All Migrant Workers and Members of Their Families* 1990 outlines extensive rights and protections for regular and irregular migrants, but this does not amount to a right to resettlement. Nonetheless, this Convention has struggled to gain the support of the international community and remains highly contentious. Overall, as migration specialist for the International Labour Office, Patrick Taran, reminds us, 'there is no international standard to recognise and measure the protection needs for people fleeing generalised civil disorder, environmental devastation or economic collapse that threaten human survival' (2001: 13).

Does this dire scenario point to the failure of human rights paradigms in both theory *and* practice? Are rights simply too individualistic, too stained by the interests of the powerful, to provide a viable ethical grounding? Poststructuralist ethical approaches to 'rights' differ radically, with some theorists taking the view that 'rights' are irrevocably tied to the model of the discrete sovereign subject (the liberal humanist individual) and as such cannot articulate an ethics of the Other (Brown 1995).[10] In spite of this, inspired by Levinas, I would suggest otherwise. Levinas asks that we reorient our understanding of human rights away from the contemporary focus on individual freedom and towards the Other (Burggraeve 2005). In his piece 'The Rights of Man and Good Will', Levinas puts forward the thesis that human rights must be concerned 'above all' with the 'right of the other man' (1998: 158). Moreover, he addresses the struggle for what he knows are 'difficult rights', noting that:

> It is not always easy, in defence of the rights of man (and this is an important, but practical problem) to establish an order of priority for those concrete rights. It may vary as a function of the actual situation in each country. (Levinas 1998: 156)

The challenges of implementing a concrete right (the issue of the 'Third' or justice, in Levinas) does not negate the role of human rights as the 'surpassing of the political, or stronger still as the critical questioning of every socio-political order and rationality' (Burggraeve 2005: 80; 2006). In other words, within a Levinasian framework human rights precede and exceed the law and politics and are the force of an irrepressible critique of current structures (Ziarek 2001: 69). Simultaneous with the enactment of any politics is the ethical encounter and the imperative to offer hospitality that this entails. We are not discrete individuals with entirely autonomous capacities. We are interdependent upon the social and ecological world we are born into. We yearn for its welcome to extend to us and offer accommodation, and in turn we are compelled to negotiate the condition of sharing this dwelling place with others. Writing from a Levinasian perspective, Ewa Ziarek identifies the impulse towards responsibility underlying all human interactions:

just as the asymmetrical relation to the Other from the start opens the question of politics, so too the proliferation of differences characteristic of democratic politics retains a reference to anarchic responsibility, which situates the subjects participating in a democratic process as accountable to others. (Ziarek 2001: 66)

Facilitating the development of a right to mobility is a necessary step to take if we want to continue to make appeals to human rights principles as foundational to the moral fabric of our social world (Pécoud and de Guchteneire 2006, 2007). A human right to mobility is driven by the anarchic responsibility Ziarek names as a core component of accountability. This right must be operating alongside any policy designed in the name of CBDR paradigms of managed migration.

AN OTHER-ORIENTED PROTECTION AGENDA FOR CLIMATE MIGRANTS

In the last two sections I have outlined alternative expressions of responsibility in relation to climate change adaptation. The first was CBDR, which places the onus on the Global North to accept climate exiles. The second emphasised the need for a human right to mobility informed by Levinasian responsibility. The primacy of the decision of the sovereign state is retained when we focus on managed migration programmes (as CBDR-informed models tend to do). The sovereign state determines who enters and how they arrive. It keeps borders stable and unperforated. Even as this is profoundly problematic, in response we cannot simply abandon all borders. Abandonment is not the solution; negotiation of borders is what must be undertaken (Bulley 2006). To predict and prepare for climate migrants is crucial; as Nick Mansfield contends, 'appeals to hyperbolic generosity amount to nothing [and indeed we] need a plan' (2012). Or, in Derrida's words:

> We are not dreamers . . . we know that no government, no nation-state, will simply open the borders, and in good faith we know that we don't want to do that ourselves. We would not simply leave the house with no doors, no keys and so on and so forth. We protect ourselves, OK? Who could deny this in good faith? But we have the desire for this perfectibility, and this desire is regulated by the infinite pole of pure hospitality. (Derrida 2001b)[11]

Managed migration is clearly an organised and planned response. In order to abide by an Other-oriented relational ethics this mode of migration planning requires that we remain vigilant. We must pay close attention to the conditions placed on the person migrating as well as the actions of the host country in protecting the migrant and reducing opportunities for exploitation. This is the 'desire' for 'perfectibility ... regulated by the infinite pole of pure hospitality', that Derrida speaks of.

However, Other-oriented, managed migration is only a partial response to our ethical dilemma. The reality is that people move in ways that we cannot always regulate. Thus, people may arrive at the world's shorelines or land borders without 'permission'. That is, without a valid visa. Part and parcel of all historical periods are practices of human movement that 'leak' outside of the parameters of acceptability or legality (Nail 2015: 26–7). There is unknowability *inherent* to this condition. The most profound political challenge we are faced with is to encounter this condition of unknowability without freezing up, closing in, or acting aggressively against it in an attempt to control what can happen. The moment we deploy a tactic of closure, we turn our back on the Other and fail to dwell-with. We sever our relational bonds and assert our own entitlement to place and identity over another's. The response-ability to assist the Other is ingrained in the human being. To cultivate an ethics of dwelling-with the Other we must advocate for a protection agenda for irregular climate migrants.

I have been arguing that the ethical injunction adaptation to climate change imposes upon us is to reflect on our dwelling and to respond to the Other. Within a Levinasian framework, ethics is realised in the encounter with the Other. This encounter always comes with a demand to extend hospitality. It is now necessary to consider how this ethics meets the historical, political and social realities of our era. We find ourselves unable to fulfil the promise of unconditional caring for Others. We are confronted with the conditional nature of all our responses: all hospitality, all gestures of welcome, all expressions of care, are conditional. On the flip side, sovereignty too, is a historically contingent concept and institutional setting, aptly likened to a 'sponge' by political theorist Jens Bartelson (1995: 15). That is, it is not unconditional by nature.

Taking Levinas's ethical theory seriously, we must accept that we are 'hostage' to the demands of the Other (1969). This means that we give all that we have; we release any grip on our possessions, particularly on the notion of an exclusive dwelling, and express generosity and care towards the Other. In practice, this is impossible, and Levinas knows this. The relational ethics I have outlined must operate in diverse political contexts where land, place and belonging remain contested and culturally complex issues. At this point, that is, immediately, Levinas introduces what he calls the 'Third', or the question of justice (1969, 1981). The question of justice comes about because we are always responding to more than one Other. There are many Others and that is why it is impossible to enact unconditional responsibility. The face-to-face relationship between the self and the Other is characterised by the infinite responsibility I have towards that person. The emergence of the Third signals the reality of limits and of decision-making that must negotiate competing demands. A crucial point in this is that our relationship with all of these Others is grounded in the original ethical relationship: thus our actions must be faithful to an overriding concern for all Others.

The emergence of the Third is immediate in Levinas as the 'stranger poses a question in both the ontological and the ethical realms' (Greisch 2011: 216). While the ontological and the metaphysical, or the political and the ethical, are distinct domains of analysis and concern in one sense, in another they tie in with each other in a way that cannot be untangled: 'alterity is not just "infinitely transcendent" and "infinitely foreign" but both incomparable and comparable, transcendent and inscribed in immanence' (Ziarek 2001: 66).[12] No duality truly exists. Levinas's model of justice, then, is inspired by the ethical imperative to respond to the summons of the Other because his life (and by extension death) concerns me; it is my business.

What would it look like to draw upon this ethical theory to develop a protection agenda for 'climate migrants' or 'climate refugees'? 'Protection' in international law is the most forceful ethical principle we have available at present. It is concerned with the overriding vulnerability of the victim. As I have described, at an international level, normative gaps exist with reference to the protection of people who cross borders as a consequence of climate change. These gaps exist in a context where there is already a dearth of rights associated with mobility. When we speak of 'protection', we are referring to two key provisions. Firstly, rights, and secondly, humanitarian needs (Kälin and Schrepfer 2012: 17). It is worth repeating that in contemporary articulations of rights, even human rights law cannot guarantee the full protection of mobile people: it cannot promise the right to refuge in another territory for those on the move, even as 'exit options' from one's homeland are available (Pécoud and de Guchteneire 2006: 75). To bring Levinas into this conversation we would need to emphasise responsibility as the foundation for any articulation of rights. Rights are useless without this grounding. Responsibility comes first.

There are two primary political and legal obstacles to this becoming a reality. Firstly, because we begin the conversation with a legacy of 'rights' that impresses the import of self-interest we are stuck with the understanding that one either has or does not have the 'right' to be in a specific nation-state. Thus, the debate that emerges is whether a person is entitled to be on the 'foreign soil' and for how long. If we were to start this conversation with 'responsibility', the power dynamic between the 'citizen' and the 'foreigner' or the 'host' and the 'guest' would look quite different. I will develop this below. The second problem which hinders the application of effective protection regimes is the complexity of the distinction between forced and voluntary movement related to climate change (Kälin and Schrepfer 2012: 2).

Thinking in terms of what rights an individual has in relation to a particular piece of the earth is a symptom of dualistic thinking. It relies upon the sharp distinction between you and I, mine and yours, citizen and foreigner. It is also the effect of building our political theories on the groundwork of stasis and the figure of the citizen, over and above movement and the migrant. For example, in late 2016, the European Union signed a deal with the government of Afghanistan. This migration deal allows European countries to forcibly expel as many 'rejected' Afghan asylum seekers as they want to. These people will be returned to Afghanistan against their will. Reportedly, some Afghani migrants had been living and working in Germany for as many as seven years before they received notice of their immediate and legally binding deportation (Rasmussen 2016). A protection system that reiterates these dualisms is always, to some extent, going to keep the 'foreigner' in a position of subordination or vulnerability in respect of the decisions of the 'host', even as it emphasises a responsibility of protection on the part of that host.

The question 'where?' has the potential to call into question the contemporary emphasis on allowing the nation-state to choose when and how people seek refuge. Where can you go? Where do you live? The one on the move is presented with these questions. Confronted with such questioning, fear and panic is carried in the sound of their response and washes up on our shorelines, over and again. Nonetheless the question 'where?' contains the possibility of calling all of us into account. Yet when this question is uttered today the tendency is to react with further interrogation of the foreigner. Instead of reflecting on ourselves, the migrant is asked many more questions. The response at present is: 'Well, let's see if you can go home.' A 'returnability test' is applied, using three criteria: permissible, feasible and reasonable. Would you face persecution or loss of life upon return (permissibility)? Is return practically possible, that is, are there any administrative or technical obstacles such as flooding or lost passports that might prohibit your return (feasibility)? Will your home government provide 'assistance or protection' upon your return (reasonableness)? (Kälin and Schrepfer 2012: 65–6).

Where can I go? Another response we might hear today is: 'Have you been forced to flee or voluntarily left?' Kälin and Schrepfer contend that this distinction is both necessary and vital as there is a prerequisite to tier 'need' and the corresponding distribution of resources (2012: 62). However, environmental policy experts Biermann and Boas rightly point out that 'adopting voluntariness as a defining criterion [of movement] would be either analytically not useful or morally dubious' (2010: 65). It is analytically problematic because 'choice' is too complex for clear distinctions to be made. Even if one's homeland has temporarily recovered from the cyclone that ripped up the housing and obliterated the economic base of the region, even if life would 'go on' and the government could promise to try and offer some of its scarce resources to the displaced and homeless, would this constitute an 'adequate' home? Are environmental and economic deprivation enough to constitute a right to refuge? The narrow application of protection, particularly refugee protection, has been challenged by migration specialists and legal scholars. Patrick Taran has rightly illuminated the 'increasingly difficult' task of maintaining a binary between the violation of political and civil rights and those of economic, social and cultural rights (2001: 29). While Taran is not sure where these protection needs should be accommodated, legal scholar Michelle Foster has developed a thorough argument in favour of an interpretation of the Refugee Convention 'capable of encompassing claims based on economic destitution' (2007: 1). Jane McAdam provides a useful intervention on this point, noting that:

social or economic deprivations can amount to persecution [as per Foster's re-interpretation], [and thus] it is not enough simply to show the 'sad fact of life' that poverty is widespread; [yet] it remains necessary to demonstrate a discriminatory element in how rights are restricted or withheld by (or with the acquiescence of) the state. (McAdam 2009: 588)

A Levinasian ethics cannot provide the normative infrastructure needed to be operative in the international arena today. This is not its purpose. Normative legal agendas need to negotiate the complex work of categorisation as well as draw clear lines between forced and voluntary movement (Kälin and Schrepfer 2012: 29; see Foster 2007 for a critical review). This is done because we live in a world divided into nation-states; and without developing agreed-upon categories and definitions, states invariably opt out of ratifying international law. Levinas's theory, when taken literally, undoes the authority of the state-as-host to pick and choose to whom it will respond. The sovereign, in Levinas, is summoned rather than choosing to turn up when it suits them. Nonetheless, this philosophical orientation can contribute to the alteration of normative agendas by continually, unrelentingly, emphasising the priority of the Other.

The force of Levinas's philosophy reads as completely unreasonable in today's increasingly protectionist political context. But we must recall, as Schroeder does, that 'ethics is not a matter of reasonable expectations' (2013: 479). Indeed, I would go further and suggest that what we currently determine to be 'reasonable' or 'pragmatic' is constructed by the dominant politico-institutional frameworks that we are operating within, that is, the sovereign state. To prioritise movement and the migrant amounts to a political theory that deconstructs the nation-state, or sovereignty. Protection agendas offer this, which is precisely why they struggle to gain much support in today's global arena.[13]

MOBILITY AS SOCIAL TRANSFORMATION

How a political community responds to mobile peoples tells us an enormous amount about the social and cultural values and practices of that particular historical period. This is perhaps why Nail reminds his reader that the story of migration is never simply individual or personal in nature; mobility is about the 'transformation of society' (2015: 13).

What he means by this is significant: we make sense of space and time via an analysis of the organisation of social movement. Human movement shapes historical periods and alters the configuration of territories and communities. Mobility forces change in the realms of politics, law and economics. Mobility tests the ethical values that different social and cultural groups claim to hold. The ideals that the United Nations advocates, broadly speaking, include those of human dignity, co-operation, safety, opportunity, protection and refuge, amongst many others. It is these values that are stretched when faced with events that signify as a social or historical turning point.

For Nail, what we witness time and again are enormous efforts to manage and contain the socially transformative potential of mobile peoples. This is achieved through one of two political techniques of power pivotal to the nation-state. Firstly, expulsion of the migrant or 'illegal' person from sovereign territories, and secondly, normalisation of the migrant into the economic system of the host society through labour programmes and tourist visas (Nail 2015: 14). Thus, we either expel or normalise. The application of CBDR to frame labour and humanitarian managed migration is an example of this process of normalising movement.

Despite political and legal efforts at containment and regulated organisation, every historical period is marked by forms of mobility that 'leak' outside of the socially derived boundaries of the era (Nail 2015: 26–7). Consequently, sovereignty has needed to split off; to stretch out and into other territories in an effort to control migration from afar. As geographer Alison Mountz explains, 'border enforcement has grown more dispersed and sovereign practices more transnational' (2011: 120). The reality of these 'leaks' is why we need to go further than managing migration. We need also to develop protection agendas and categories of refuge for environmental exiles and migrants or, even – if we can utter such a contentious label – 'climate refugees'.

In relation to climate change, efforts to contain and regulate, normalise and control are under way, but confront a startling reality: dangerous climate change is so big and so unpredictable that the possibility of extensive and mass instances of human mobility constitutes more than a leak. Climate mobility may point to a rupture that brings to light the faulty mechanics of our migration systems, predicated as they are on the primacy of citizens and states. Following WWII, Arendt was adamant that national sovereignty was completely incompatible with respect for human dignity (Isaac 1996: 61). Indeed, Arendt went so far as to call for new forms of political community and action. Without the creation of alternative political domains, the non-citizen, the rightless, the dispossessed, expelled and unwanted would be relegated to the 'internment camp, a site of prolonged homelessness, an institutionalised limbo' (1996: 63). The nation-state cannot, for Arendt, secure rights and dignity. However, philosopher Michelle Boulous Walker summarises an important distinction found in theoretical approaches to institutions such as the nation-state: the instituted order and the instituting drive (2017: 6). Sociologist René Lourau elaborates: 'What I and others before me have called the *instituted*, the established order, the already existing norms, the state of fact thereby being confounded with the state of right. By contrast, the *instituting* aspect . . . has been increasingly obscured' (cited in Boulous Walker 2017: 6).

Derrida engages with this distinction between the instituted and the instituting when he discusses the relationship between law and justice, unconditional and conditional hospitality, and unconditional and conditional sovereignty. Within this logic, unlike Arendt, Derrida is able to articulate an understanding of state-based sovereignty that is always already open to its own demise, rearticulation and reformation. Arendt may want to do away with sovereignty entirely, but within a deconstructive thinking this is impossible. What we have instead is constant interrogation in the name of justice. Mansfield succinctly highlights the significance of a deconstructive approach to state sovereignty when he writes that our task is not to relinquish sovereignty but to take it up as an act of responsibility:

> This taking responsibility is a kind of sovereignty . . . it is an unstable authority, instituted in the events that both enact and open the question of its legitimacy. It is a collective authority but one always open to being questioned. That sounds like government . . . It is a call to recover the distinctive and active responsibility of government itself. (Mansfield 2012)

Ruptures produce crises. Rather than falling into regressive forms of nationalism, these ruptures invite re-evaluation of the status quo. This should not mean a confirmation of outdated ways of operating. When the subject or sovereign is faced with a crisis, what emerges is never an undisturbed self-same identity. In this way, against Arendt's assertion, we can press for the recognition of dignity and security within the state structure, but only if this very structure is willing to remain open to that which calls its fundamental existence into question.

The language of rupture risks further invoking fear-laden discourses of migration that characterise national debates around the world. Yet this tearing, this break with the present modes of organisation, will occur regardless of any nationalist spells of angst and anxiety. Already today, chronic, mass and sudden mobility related to disasters, internal conflict and political instability is poorly regulated by the international community (Martin et al. 2014: 124). Failing to provide any legal and material infrastructure does not negate the reality of displacement. Nor does it stop people from fleeing dangerous situations or seeking out more secure livelihood options. The social transformations driven by migration are under way regardless of the Global North's collective denial and refusal to engage with this reality.[14] The task is to meet this challenge as a possibility rather than a threat. Susan Martin points out that 'frameworks support or constrain mobility'; adequate structures can provide the conditions of basic safety for vulnerable peoples travelling, while the absence of services increases the experience of precariousness and violence (2012: 1052).

To turn towards the migrant would entail consideration of how to 'strategically re-organise movement' (Nail 2015: 14) in the interests of the vulnerable, rather than the nation-state. Calls have been made by the IOM to institute 'dignified, orderly and humane' migration programmes in response to climate change impacts (Lacy Swing 2015: 79).[15] We may not be able to predict with certainty the 'implications of environmental change for migration' (Bardsley and Hugo 2010: 238), but we do know that it happens. Articulating 'new' categories of movement and 'new' policies for managing mobility are the work of politics, but more than this it involves the negotiation of foundational ethical concepts and norms. It is the effort of taking responsibility. The strategic reorganisation of movement can, of course, be devoid of ethical consideration: it can be aggressively self-interested, as in the case of the containment of asylum seekers in internment camps or prisons throughout the world.

CONCLUSION: CONTESTED GROUNDS

So far, I have demonstrated that there is a definitive preference for facilitated or managed migration or relocation policies as part of any future adaptation platform including human mobility. There are two primary reasons for this. In the first instance, orderly migration does not challenge the territorial borders of the nation-state. Secondly, regulated migration has a greater chance of ensuring the safety and dignity of the migrant. Consequently, pre-existing migration programmes, or one-off exceptional migration measures granted by a state in accordance, for instance, with fulfilling one's CBDR, keep the authority of the sovereign state intact.

It is important to return again to Derrida's question 'where?' because it asks us to look very closely at what is going on in the enactment of hospitality towards the stranger. In the examples of regulated migration or fulfilment of state duties, the following scenario is played out: the host (the nation-state) welcomes a select number of climate migrants or climate displacees. Next, the host swiftly and firmly shuts the door behind the guests. This exchange does not sufficiently understand Derrida's point. The power of the question 'where?' extends beyond attaining welcome for some. Its power resides in its capacity to call the very existence of the sovereign into question. To force a change in the institution of sovereignty itself. This is far more radical than coaxing states to invite some additional people in. It calls upon the sovereign to recognise its own contingency and inherent porosity. In doing so, the question 'where?' actually requires us to understand that we are not sealed off from the world; we are not separate. Our very language of citizen and foreigner, us and them, inside and outside, perpetuates the delusion of permanent dualities. We are interdependent. Acknowledging our interdependence carries with it a spontaneous acceptance that we must care for one another as a condition of being human on this earth. After all, 'no border is guaranteed, inside or out' (Derrida 2011: 105). The self and the sovereign state must always meet the demands of the foreigner. It can never be free of this requirement.

PART B

Chapter 4

IN AND OUT OF PLACE:
THE CASE OF BANGLADESH

'In the most densely populated delta in the world it is not possible to differentiate between land and river, human populations, sedimentation, gas, grains and forests, politics and markets. Human habitations are superimposed on an even more dense river system, which is a constantly shifting, soggy planet.'

Nabil Ahmed[1]

WELCOME

We arrive very late, close to midnight. In the car, we move through crowds of people, dodging colourful rickshaws and damaged trucks, weaving around pedestrians still wandering the streets. Some small groups of women huddle in corners, protecting one another. We turn another corner and the people are fewer. Here, armed guards patrol buildings with large fences. Some of the fences have barbed wire running along the top; others have shards of glass jutting out. Large, new-looking commercial and private dwellings dominate the landscape. Our driver is taking us to our hotel, situated in the wealthy diplomatic zone of Gulshan. Later I learn that this area, along with its neighbouring region, Banani, is where all the foreigners stay.

Dhaka is a city under construction. Old buildings are torn down by hand. The labouring bodies of the men continue into the early hours of the morning, with the smacking sound of sledgehammers and axes hitting concrete interrupted only by the dawn prayer. The broken rubble forms small hills on the sides of the road, resources for other sites now, while the cleared space makes way for a new development. Many of these men have journeyed from afar, making the increasingly common migration from the rural areas to the urban centres.

I am welcomed here in Dhaka, in this place where these labouring bodies work day in and day out. This welcome shields an underlying violence. Slums dwellers are routinely evicted from the city of Dhaka to its outskirts, all in the name of 'development'. Labourers are underpaid and live precarious lives, with hotel staff better off than those who build the hotel, but also exploited. I am welcomed in this place. I am fed well for little money and provided the comfort of a new and stylish boutique hotel room. Even when illness hits with a force I have never experienced before, I am well attended to, cared for, fed and given the space to recover.

I am here to talk with people about climate change and its implications for issues of mobility. Researchers are doing this more and more and I realise that some of those I am interviewing have done this work of talking with foreigners over and again. They have routinely offered a professional hospitality. While we start with migration, inevitably conversations turn towards issues of land ownership and rights, housing, poverty, government corruption, and international responsibility. What I come to learn is that it is land ownership, and the related but distinct issue of housing or shelter, which takes centre stage in Bangladesh. But now I am jumping ahead of myself.

This book advocates for a relational ethics of dwelling, and aims to facilitate political action that embraces our responsibility (for the Other) to create and sustain safety, opportunity, hospitality, protection and care. There is an urgent need to articulate the fundamental right to a home as a material foundation, and concomitant rights to a social texture and livelihood, both of which are conditions that exceed the basic premise of a roof over one's head. Levinas emphasises the condition of dwelling as pivotal to human experience when he refers to it as the 'condition ... [the] commencement' of 'human activity' (1969: 152):

> Man abides in the world as having come to it from a private domain, from being at home with himself, to which at each moment he can retire. He does not come to it from an intersideral [empty/void] space where he would already be in possession of himself and from which at each moment he would have to recommence a perilous landing. (Levinas 1969: 152)

In other words, we are not born into the world already in possession of a separate 'human subjectivity' and dwelling. Dwelling is not the mere announcement of our arrival into the world; simply being-here. For Levinas, dwelling involves an act of possession that is necessary in order to take 'refuge empirically in the home' (1969: 154). Yet as I explored in Chapter 3, at the very moment of possession there is a simultaneous contestation of my control and my enclosure in the home. The absolute Other, the stranger, 'paralyses possession' and indeed compels the one at-home to welcome the Other into their home (1969: 171). So significant is this disruption of one's dwelling that Levinas declares that the primordial openness to the stranger is as 'essential to the essence of the home as closed doors and windows' (1969: 173). Thus we have a complex dance between being-at-home as a condition of worldly existence, and the primordial interruption of this by the stranger, simultaneous and necessary.

We have seen in Chapter 3 that through Levinas we can shift human rights away from a single-minded pursuit of individual interests (egoism) and towards the rights of the Other (Burggraeve 2005: 95; 2006; Levinas 1998: 155–8; Ziarek 2001). Such a reorientation is crucial to return to over and again in relation to the issue of dwelling, where all too often egoistic interests prevail, with the temptation to lock the doors and close the curtains frequently given in to. Levinas argues that while being-at-home is necessary to the 'goal of need', the capacity of the Other to challenge this points to the manner in which 'possession itself refers to more profound metaphysical relations' (1969: 160–2). That is, we do not possess a dwelling in the same way that we may possess any other object in the world. The dwelling occupies a special place. It is a place where the face of the Other demands to be heard, and where ethics as hospitality must emerge. One's dwelling permits the hospitality evident when you extend welcome to another, but also the welcome one receives daily when returning to one's private refuge (Levinas 1969). It is for this reason that land, housing and property rights must be informed by this ethics of the Other and take centre stage in climate justice.

The importance of an Other-oriented paradigm of human rights is apparent when we unpack the issues of land, housing and property at play in Bangladeshi society. This chapter will first examine the politics of dwelling as a relationship to place in Bangladeshi adaptation discourse. It is land that is at stake, vulnerable to the forces of the climate system and subject to the power dynamics of society. Consequently, successful climate-adaptation policies must engage with calls for land justice in order to move towards securing housing, land and property (HLP) rights.[2] HLP charters call for a human rights approach to this area of concern, with policy advocate Scott Leckie noting that HLP rights have had an inconsistent history within the UN despite the crucial role they play in the restoration of lives and communities post-disaster (Leckie and Huggins 2011). At the domestic level, the local Bangladeshi organisation, the Association for Climate Refugees, works with the international group, Displacement Solutions, both of which name rights to housing, land and property as foundational to all other rights claims. In their view, we cannot achieve safety and protection without some form of 'adequate' shelter or housing (Displacement Solutions 2012, 2013).

At a practical level, however, there is only so much land, with the IOM warning that within Bangladesh there is very limited scope of inland migration (Walsham 2010). Local adaptation meets an environmental limit. This brings to the fore the practical need for policies related to climate mobility. In the second half of this chapter, I will argue that the social phenomenon of migration invites us to engage with the responsibilities of hospitality, that is, of welcoming the foreigner. Clearly, mobility will need to be recognised as a climate impact requiring an international response due, in the first instance, to mass migrations internal to Bangladesh and, secondly, because of the likely scenario of substantial land loss to the country's territory. Once again, it is land that is pivotal. With mobility rights comes a challenge to others to share their lands; to open their dwelling to another. The question of the foreigner, 'where?', calls my entitlement to place into question and indeed 'puts in question the world possessed' by 'me' (Levinas 1969: 173).

What Derrida terms 'structures of welcoming' (2002: 361) must be developed both in advance and in response to climate disasters for this politics to operate in the spirit of a relational ethics of the Other. In such performances of hospitality there is the possibility of creating new homes for those left without a means for survival, sustenance and livelihood. As important as they are, measures that intervene only at the point of disaster recovery offer only a partial response. A relational ethics can only be achieved if we can hold together the inherent tension between place and placelessness identified earlier as integral to dwelling: fostering policies that provide homes to those rendered homeless while questioning the grip of possession.

LOCAL ADAPTATION: SITUATING CLIMATE CHANGE

Bangladesh emerges out of the lands that interweave with the Ganges Delta, making it an area with fertile and nourishing grounds as well as a volatility that threatens to render the region uninhabitable. These low-lying deltas constitute the means of survival for Bangladeshi people as well as their exposure to extensive seasonal flooding. This context reveals the dialectic of nature itself, of creation and demise, of life and death, and of the inextricability of the grounds of our nourishment and our extinction. The Greek word *pharmakon* comes to mind. Within a Derridean lexicon, *pharmakon* captures the paradoxical nature of something containing 'alternatively and simultaneously' that which is 'beneficent or maleficent' at its heart (1981: 75).

Historically, Bangladesh has dealt with a diverse range of climatic events. In addition to flooding, it has weathered intense cyclones and prolonged drought. Between 1970 and 1998, the country experienced no fewer than 170 major disasters (United Nations Habitat 2011: 2). In other words, Bangladesh is no stranger to difficult environmental conditions. As a result, local and aboriginal adaptation techniques, as well as the development of large-scale disaster management programmes, have been employed. With the increased frequency and severity of weather events associated with anthropogenic climate change, these modes of adaptive capacity and local resilience are placed under greater strain, while forms of damage such as river erosion and sea level rise are impacting coastal regions in irreversible ways (Akter 2009). This irreversibility has witnessed the predominance of the language of 'Loss and Damages' in local development organisations which are now thinking through the topic of compensation (not merely voluntary financial contributions) for the effects of climate change.[3]

Constant and challenging climatic conditions have contributed to the forces of unplanned and massive urbanisation in the last few decades, resulting in the city of Dhaka becoming the tenth most populous in the world, subsequently attaining the status of a 'megacity' with around 15 million occupants. Since national independence in 1971, two-thirds of all urban growth can be traced back to migration (Afsar 2003: i). The *Bangladesh Climate Change Strategy and Action Plan 2009* (hereinafter 'Action Plan 2009') indicates that the government expects this figure to reach 40 million by 2050 (Ministry of the Environment and Forests 2009). Anecdotally, these migrants largely come from 'environmentally vulnerable regions', though methodological obstacles make this difficult to confirm empirically (Walsham 2010: xiii). As I have discussed in Chapters 1 and 2, gaps in the empirical evidence have resulted in delays and even the refusal to develop specific protection and rights regimes for climate migrants or 'refugees'. 'Climate refugees' is a category that has no legal basis, but circulates as an important cultural and social discourse in Bangladesh. For instance, in Bangladesh's leading English language newspaper, *The Daily Star*, stories of 'climate refugees' moving into slum housing in major cities are not infrequent.[4]

A highly active and effective civil society comprised of international and domestic NGOs, political activists, research organisations and economic enterprises such as Building Resources across Communities (BRAC) have ensured that since independence, the people of Bangladesh have not been left behind in social and economic policy decisions.[5] Defined by the international community as an LDC, Bangladesh has been able to access special adaptation financing schemes reserved for the most vulnerable. The role of this civil society in reducing the prevalence of corruption and ensuring some degree of transparency has been significant.

Part of the utilisation of adaptation funds by Bangladesh has been the activation of the development sector in responding to the challenges of climate change. The mobilisation of development resources and discourses conforms to the climate–development nexus discussed in Chapter 1. With 63 per cent of its citizens dependent upon employment in the agricultural sectors, development work has assisted with adaptation projects focused on farming, fishing and other local practices (Kartiki 2011: 25). With funding provided by the Bangladesh Climate Change Resilience Fund, the government, in conjunction with a local organisation, has been able to implement a community-based afforestation and reforestation project which aims to build local resilience. Forest coverage helps to protect communities when storms and cyclonic activity arise. Moreover, necessary resources have been channelled into the ongoing development and construction of early warning systems and cyclone shelters (Bangladesh Climate Change Resilience Fund 2016).

Climate adaptation aligns well with a localised development agenda. However, rather than focus on the many ways that this operates, this section argues for a relational ethics of dwelling that is materialised in the form of HLP rights. I add my support to the call for justice which understands the pivotal role of such rights for any effective climate adaptation. Land is an asset that is increasingly difficult to guarantee with the forces of neoliberal capitalism, corrupt local power officials and individuals, rising sea levels and turbulent weather.

A POLITICS OF PLACE: LAND AS LIFELINE

In the context of Bangladesh, we need to appreciate the centrality of land to life, as both the basis of survival as well as enabling political influence and social capital (Feldman and Geisler 2012). The importance of land differs in the rural areas compared with the urban contexts. In rural regions, land is crucial to livelihood, with the cultivation of food and produce for sale. In urban areas, ownership is less necessary, but housing rights and security of shelter paramount (Gelbspan and Thea 2013). The relationship of land to survival is at the most basic level a reference to its role in the cultivation of produce with, as noted above, over 60 per cent of Bangladeshi people earning a living from agriculture (Kartiki 2011: 25; Walsham 2010: 23). Adaptation policies must adequately negotiate the local politics of land in order to offer truly sustainable internal solutions. This is because, according to Barkat et al., 'Land forms the basis of our [Bangladeshi] social, economic and political power structures' (2000: 1).

More than just economic or political, land ownership has social significance. When I interviewed the founder and CEO of the Bangladeshi NGO, ACR, Muhammad Abu Musa, he noted that land offers a specific social status in Bangladeshi society. Without land 'you are nothing', he said.[6] It is perhaps for this reason that the Bangladeshi Constitution had the foresight to inscribe property rights for all citizens into its Charter. Yet despite this ambition, half of all rural Bangladeshi people are landless or in possession of only half an acre. In urban regions, the minimum condition of shelter or housing is needed in order to offer safety. Land corruption in Bangladesh makes confronting the issues of land rights and associated (though distinct) rights to housing and property, of paramount importance for providing sustainable internal solutions with regards to climate change adaptation.

Land is bound up in complex and contested political, social and economic relations. The two most contested forms of land in Bangladesh are *khas* lands and *char* lands. *Khas* refers to state-owned land. This land is often at the margins of society such as on the coasts and the river deltas. In law, this land is to be redistributed to the landless through rehabilitation and resettlement programmes. However, there are numerous obstacles to this becoming a reality. In addition to the power held by the wealthy, there are social norms that continue to prohibit land ownership. It is evident that particular groups of landless peoples are excluded from the possibility of ownership 'such as those headed by unmarried women or widows with only daughters or no children at all' (United Nations Habitat 2011: 7). The dynamic geography of the region compounds this experience of injustice.

Char lands are inherently transient. They are formed through the accretion of soil, which means they are also highly susceptible to eroding. When these lands wear away and completely disappear, it leaves dwellers without a home or land with which to rebuild or cultivate foods for consumption and profit. However, this natural process also means that new lands emerge regularly. The new land is quickly grabbed by the powerful members of society. Soon after Independence, in 1972, a Presidential Order was passed which renamed all *chars* as *khas* lands. The Order stated that the government would distribute the *chars* to the poorer farmers (Feldman and Geisler 2012: 978). However, what has unfolded is a 'radically unstable' system of land tenure, where the wealthy are able to wield their power, dispossessing the poor (2012: 978).

In Chapter 1, I argued that adaptation is primarily about dwelling as relationship to place: it is about how we value place, whose place matters, who has the right to a place and what our duties are towards our own place, as well as that of others. According to Sajid Raiham, a senior advisor with ActionAid Bangladesh, the systems operating to deal with land transferral (following erosion and the emergence of new lands) perpetuate a systematic denial of rights for the poor and landless:

Sajid Raiham: The laws relating to the *khas* land and *char* land are not very respected in our country. There is a systematic denial – the systems are supposed to . . . So, you have land in this particular area, okay, but it is unavailable so you become landless, so you shift from here to there, outside of the riverbank. After 50 years, the new land is there [the new *chars* form]. You had land there and you know that after 50 years it is not yours any more, it is your son's or your grandson's. They know that their forefathers

had the land there okay. But they didn't have the proper papers off the government, they didn't have proper records. So because of the system, these people, the landless people, though they have land they cannot claim it. What happens then is that the powerful people take it.
Elaine Kelly: So what do they do with that land?
Sajid Raiham: They lend that land to these people, the landless people. Now, the system is that they do not take money from the landless people. They say that you will be cultivating the land and that you have to give me 50 per cent of your earnings. So these people, the landowners, become kind of lifesavers for the landless because they have no place to go.[7]

This can continue to occur, I am informed, because Bangladesh has no formal relocation policies. A secure place for the poor is difficult to attain. In the case of river bank erosion and displacement, land redistribution programmes designed to provide restitution for victims in Bangladesh 'lack transparency and so priority is given to river bank erosion victims who were previously larger landowners' (Zetter 2011: 44–7).

An implication of this lack of transparency regarding land redistribution is that the rural displaced are left without land and consequently a home and a livelihood, forcing a decision on whether or not to migrate.[8] Should they migrate, the journey is dangerous and difficult. I was told of people using a plastic sheet as shelter from the weather as they made their way on foot to the city, taking rest on the side of the road. Alternatively, if remaining in their original community, people find themselves at the mercy of the landowners. It is these complex economic, political and social conditions which point to the practical and moral impossibility of cleanly distinguishing between forms of forced and voluntary migration.

When Arendt noted that the loss of home equates with a loss of the 'familiarity of everyday life' she was pointing to the profound implications such a loss can produce, especially regarding the capacity to imagine a future (1996: 110). Forms of structural neglect produce experiences of extreme marginalisation. This is particularly the case when people are dispossessed from their lands, whether through human actors or natural conditions, or more likely, a combination of the two. Giroux has referred to this process as rendering certain populations 'human waste' (2006). Giroux's language is brutal and confronting. The notion that in today's self-reflexive and interconnected world we could reduce a human life to waste is frightening. However, the condition of mass internal displacement brings into sharp focus the question of whose place we are seeking to protect, restore or rehabilitate, and whose we are leaving to ruin; the informal burial grounds of the poor.

The impacts of climate destruction are mediated by economic and political power relations that perpetuate conditions of social neglect. This is undeniable when we hear the following sorts of testimony from victims of climate change:

> We had sufficient cultivable lands. My father has a departmental shop in the market. We had lost all in the riverbank erosion. . . . One night, I had lost my brother and sister in a storm surge. . . . At last, I resettled in this river strip. I have been living here since 15 years with my four children and wife, but the government has not allocated any land. Fishing is about to close for frequent cautionary signal of weather. My family members are starving. . . . Sometimes, I wish to commit suicide. This settlement is also to be grabbed by the river. Where will we go then? (Cited in Nandy and Mehedi 2010: 13)

'Where will we go then?' These words were spoken in 2009 by a man living in Boyar Char, a river strip of Noakhali, a District in the South-East of Bangladesh, close to the city of Chittagong. His story is only one of many. It is estimated that approximately three-quarters of a million people were displaced in Bangladesh by river erosion between the early 1980s and early 1990s (United Nations Habitat 2011: 3). The dynamics of displacement mark this place in a variety of ways. McAdam and Saul have described it as multiple displacements: the initial displacement across a space (rural to urban) compounded by dislocation in-place once in the city, sometimes up to eight times (McAdam and Saul 2010: 251).

Some people remain in a suspended state of existence, evicted or under the threat of removal from their dwellings, without access to housing or any other shelter even as they may find themselves living in their original homelands; 'dispossession and alienation shape the current social relations of landed property' in Bangladesh (Feldman and Geisler 2012: 972). For the man quoted above, the ongoing trauma of multiple displacements brought him into confrontation with the very will to live: suicide is contemplated. The inevitability of renewed exposure to dispossession functions to foreclose the promise of a future and limits the capacity of the victim to imagine a more desirable horizon. Without structures that provide support and respond to these situations of disaster, many more people may find themselves wondering what, if anything, the future holds for them and their family. In a sense, it is not simply that those left homeless are without the 'object' of shelter. More dramatically, they are denied the hospitable condition of the empirical home: a place of retreat from the 'inclemencies of the weather', the event of welcoming itself, of recognition that suspends, if only temporarily, the 'uncertain future' (Levinas 1969: 152–8).

The institution charged with providing the social, economic and political support structures for the possibility of hopeful futures is the nation-state. This institution devolves or delegates power and resources to local and regional government and NGOs, or increasingly, private organisations. It is fundamental that the logic of human rights (of and for the Other) filters through these layers of bureaucracy and governance. While it is in the nature of *chars* to come and go, the frameworks fitted around them should be less transitive and unstable, and consequently more reliable and comprehensive for those most adversely affected. Or, at the very least, the transformative capacity of human structures should be altered in the name of addressing injustice rather than further perpetuating the conditions of disadvantage.

The Boyar Char man's narrative is disturbing and calls upon all of us to respond, not just the Bangladeshi government. For Gelbspan and Thea, land is more than a 'material good or a piece of property. . . . [For many, it is] an integral part of a life of dignity and a better, shared future' (2013: 14). Again, the notion of a 'future', of a horizon of possibility which can spur engagement in the present, is bound to the value of life. Perhaps we can say that those who are denied a future are also denied their humanity. 'Adaptation with dignity' demands the inclusion of rights to housing, land and shelter (Displacement Solutions 2012). It must encompass the possibility of a secure future embedded in these rights.[9]

THE LIMITS OF THE LOCAL

In Bangladesh, a family can find themselves needing to migrate to the national capital, Dhaka, for economic reasons (following the diminishing quality of their agricultural produce), only to be subject to the ever-present threat of removal or being pushed along to another part of the slums. Certainly, in conversations I had with locals in Dhaka in 2013, I was told of an area of informal settlements that had been stolen from the inhabitants, knocked down and redeveloped by local developers. Although the data is uncertain, many times, I was told in interviews, new residents move from the rural regions of Bangladesh to the cities because of the impact of slow-onset disaster: from the South-West coastal districts of Khulna as a result of the increased salinity and sea level rise, or from Mymensingh to the North of Dhaka, or Rangpur in the North-West where prolonged drought affects livelihood strategies. Of these people, up to 40 per cent live in informal housing or 'slum' communities when they reach the city edges (McAdam and Saul 2010).

The Bangladeshi government has officially acknowledged the role that migration does and will continue to play in response to increasing environmental and climatic pressures. The government's 'Action Plan 2008' notes that mass displacement of 'environmental refugees' from coastal regions will significantly impact the livelihoods and health of many in Bangladesh, particularly the poor (Ministry of the Environment and Forests 2008: 1). In response, little more than gestures are offered, and it is unclear how any policy will be financed. In the 'Action Plan 2009', migration and relocation take on a more significant role as a form of adaptation to climate change. The 'Action Plan 2009', notes that because over 60 per cent of the country is less than 5 m above sea level, this leaves it highly susceptible to severe flooding (Ministry of the Environment and Forests 2009: 7). While historically flooding has not resulted in considerable cross-border migration, it does contribute to internal dislocation and chronic migration to urban centres. Estimates point to the permanent displacement of up to one million people annually as a result of flooding (Zetter 2011: 14–27).

Despite the forecast of increasing rural–urban migration, the development sector continues to focus on assisting people to remain in their original homes. As a consequence, there is no funding available to support, protect or shelter people who are migrating. This lack of social services was confirmed in interviews with some development organisations. I asked interviewees if programmes to assist with safe passage from rural to urban areas, as well as housing projects for new migrants, should be supported by government. I received a mixed response. One respondent sardonically remarked that the Bangladeshi government would not want to support such measures as it was assumed that the development of urban social support (housing, welfare and so on) may increase the number of people coming to the city, something considered deeply undesirable by some. However, Bangladeshi organisations are beginning to call upon the government to support both internal resettlement and relocation policies as well as 'migration within and out the country, as well as protecting those who are most vulnerable' (Shamsuddoha et al. 2012: 9).

The issues of forced internal migration and displacement are fraught. As Zetter explains:

> For Bangladesh, the partition of India in 1947, the war of independence in 1971 and the severe famine in 1974 provide a legacy of forced displacement framed by political, social and cultural trauma. These historical experiences have inhibited adoption of policies or instruments dealing with cross-border and internal displacement. Against this backcloth, Bangladeshis have always had to cope with temporary or permanent displacement due to environmental hazards because of its low-lying topography. (Zetter 2011: 6)

Consequently, solutions to mass internal climate displacement, such as internal relocation, are complicated and haunted by historic ethnic violence associated with dispossession. The troubled region of the Chittagong Hill Tracts is a case in point. Located in the South-Eastern corner of the country, Chittagong is the site of the dispossession of the aboriginal landowners by the majority Bengali population, a situation which has created ongoing violence. While a peace arrangement was developed in 1997, the region is still subject to conflict and insecurity (Zetter 2011: 30).

Once again, land and its cultural, political, economic and social significance come to the fore. It is 'land as place', 'land as commodity', 'land as lifeline', and issues of landlessness, land loss and land grabbing with which our discourse of adaptation must grapple, especially if we want to promote the most ethical and sustainable actions available to us.

DOES THIS REQUIRE GLOBAL RESPONSIBILITY? MIGRATION AS ADAPTATION

If HLP rights are unable to be secured locally, mobility rights are also required in order to satisfactorily respond to the impacts of climate change for Bangladesh. In Chapter 2, I argued that climate-induced mobility demands a more radical expression of responsibility than what is currently envisioned by our political system of nation-states or our ethical paradigms of distributive justice. In light of this, as I discussed in Chapter 3, the enormous normative gaps that exist with regards to rights to climate-induced mobility present us with a profound crisis. In part, these gaps persist because human mobility challenges political boundaries and the authority of nation-state sovereignty. Human mobility, particularly on the scale major wars and disasters provoke, requires a relational ethics that elevates the welfare of the vulnerable persons over the interests of, at the domestic level, the elite, and at the global level, the nation-state.

The Bangladeshi government has repeatedly called upon the international community to share the responsibility for providing shelter, refuge and opportunity to its citizens. I have outlined the conditions that make this appeal to the international community, particularly the Global North, both valid and urgent. In the first instance, there are complex local land struggles and unjust land redistribution patterns that hinder local adaptation. Further, it is predicted that Bangladesh will lose land to rising sea levels and land erosion. Moreover, the scale of internal displacement and migration exceeds the capacity of the state to provide local solutions.

Bangladesh's 'Action Plan 2009' includes statements that utilise the language of 'refugees', an appeal that is ethical in nature. The government lobbies for international economic migration programmes to alleviate the stress that major cities like Dhaka are experiencing due to internal migration. Because the coastal regions will be affected dramatically enough to force the movement of people, the plan names the 'free movement of natural persons' as a justification for the 'resettlement of environmental refugees' abroad (Ministry of the Environment and Forests 2009: 2–3).

Overall it is estimated that in excess of 20 million people will be displaced 'in the near future' as a result of sea level rise, cyclones and storm surges (Ministry of the Environment and Forests 2009: 17;

McAdam and Saul 2010). In light of this, how sustainable is a call for solutions that are only internal? While the 'right to a home' must first take account of the desire for people to remain in place, there are significant political, economic, social and environmental limits to this. In the case of Bangladesh, the IOM has noted that the area is a climate 'hot spot' with 'limited adaptive capacity . . . [and] limited scope for inland migration' (Walsham 2010: 6). This physical limitation is confirmed by a United Nations report which states that 'per capita availability of cultivable land stands at a minuscule 0.09 hectares, indicating a very limited scope for any comprehensive redistribution of land' (United Nations Habitat 2011: 8).

However, the evidence to date demonstrates the unlikelihood of significant unregulated migration across national borders (McAdam and Saul 2010). This is a result of the specific ways in which slow-onset disasters, such as drought, tend to drive migration gradually. In relation to predicted land loss in coastal regions of Bangladesh, we have the benefit of foresight to prevent or lessen impacts where possible. Preventative measures can take two forms. In the first instance, there are adaptation techniques which employ technological and emergency relief solutions *in situ*. In the second instance, there is also the possibility of designing planned migration programmes which respond to predicted impacts. In relation to migration, the benefit of planning is that it may assuage some of the internal stress that Bangladesh faces in relation to rapid urbanisation.

Planned migration and relocation can be national, regional and international in scope. At the domestic level, organisations such as the ACR are developing small-scale relocation initiatives.[10] Regional tensions at the borders of India, Bangladesh and Myanmar point to one of the complications of organising cross-border migration. I will briefly touch on these tensions, as there is a strong need to develop a regional approach which recognises that migration is a form of adaptation. Paula Banerjee outlines the long histories of migration which constitute the 'frontier' regions at these regional borders. For Banerjee, while migration has taken place at these borders over many centuries, it is in the last few decades that this practice has come to be viewed as something that provokes a 'security' response from neighbouring states.

Banerjee tells us a common story, the story of human mobility as norm rather than deviation (2010: 99). Despite the naturalness of movement, the regional flow of migrants is reported in Indian newspapers as a matter of security. In these forums, this age-old practice is cast as burdensome with a 'tremendous antipathy toward migrants, particularly from Bangladesh and Myanmar' (2010: 98). Nonetheless, in the face of imposed and historically constituted borders, these communities continue to share cultural and linguistic ties. The presence of shared culture and values mean that 'some movement is inevitable' (McAdam and Saul 2010: 244). Indeed, according to McAdam and Saul, 'the Indo-Bangladeshi border remains porous, despite India's physical fencing of parts of it in recent decades' (2010: 244).

While there may be some debate regarding the validity of referring to migration as a type of adaptation, more frequently, in Bangladesh, this concept has become a mainstream view. When I interviewed Md Shamsuddoha, Chief Executive of the Centre for Participatory Research and Development, I asked about this directly, receiving the following response:

> Md Shamsuddoha: Ah yes, there is a debate in saying that migration is a form of adaptation, and there are many people saying that. Sometimes I agree that migration is a form of adaptation, and migration is a way of getting opportunities for the affected people to be relocated to other areas in the form of adaptation. It becomes clear that Bangladesh does have limits to what extent it will be able to adapt. If there are a huge number of a migrating people, displaced people – there is a prediction that by 2050 there will be perhaps 20 million people who will be migrated in this place. So if we put them under adaptation programme the Bangladesh government will not have enough resources, in terms of money, in terms of space, in terms of capacity. So yes, migration should be part of adaptation, but it needs to have some regional and international aspect. What goes with adaptation – adaptation should be, it should be within national boundaries, but if adaptation should exceed the boundaries, adaptation is to other regional areas. . . . So in that case you could say that adaptation is the process of cross-border migration to other places.[11]

If we return to Article 14(f) of the *Cancun Agreements* 2010, co-operation with regard to human mobility at national, regional and international levels is encouraged. The Advisory Group's report, *Human Mobility in the Context of Climate Change* (2015) discussed in Chapter 3, notes three avenues for addressing mobility: local adaptation measures for those displaced internally; regulated and dignified migration for those who need to move; and participatory and dignified relocation in the case of larger-scale communal relocation.

Mobility is an example of a social practice that gets stuck in political and legal processes. This is not because it is any more of a complex issue than anything else we deal with in this global world. Rather, it is fundamentally because it forces all of us to reflect on our own claim to dwell (facing the Other is constitutive of being, but we constantly deny this reality and responsibility). Consequently, like practices of local adaptation, human mobility is also about 'land as place', 'land as resource' and 'land as lifeline'. The social transformations that mobility spurs concern the demarcation of borders and boundaries as well as the conditions of belonging and ownership. This social practice asks us to share our place, our resources, our lives and our future. In other words, we must negotiate the conditions of being-at-home endlessly.

It is for this reason that migration as adaptation continues to be such a sticking point in the international arena. This highlights how a relational concept of dwelling is central to how we 'adapt' to climate change. What is irresolvable and difficult are the politics of land and place that will need to be addressed when finding adaptation solutions involving resettlement or relocation, whether internal to the state or international in scope. Further, what is up for debate is how climate mobility should be organised. In particular, who is responsible for dealing with it, and where should people go? This is especially the case in countries like Bangladesh, where it is evident that local adaptation meets serious obstacles.

Contextualising Responsibility-for-the-Other

Bangladesh does not shy away from presenting its history in all its violence and terror. I am walking now in the relentless heat of Dhaka. The streets are noisy and cramped. I feel myself pressing against bodies though at no point do I physically encounter another body. Eyes on me, all white skin (mostly covered) and long, revealed hair. I should have brought a headscarf. But I am weak and I am faint from illness, it is Ramadan and I am not sipping water in public out of respect for both cultural and religious protocols, and law. My partner, David, follows me. We find relief from the streets in a quiet museum. It is small, but packs its rooms with everything from images taken during the War of Independence and weapons utilised, to the skulls of victims. I enter a room full of enclosed bones and skulls. I am suddenly aware of my body in all its dimensions. My legs are sticking against the long cotton skirt. I am sweating, my back and arms drenching the linen top. And now I am outside of my weight, head spinning, yet completely anchored in the nausea rising from my stomach. I make a quick turn towards the exit and emerge in the light and air. I fall into a chair and take some deep breaths. I manage to avoid passing out. I look up, and there hangs Ginsberg's 'September on Jessore Road'. Our guide and his friend, a local university student, notice my interest and ask if I know the poet. I reply that, 'Yes, I love Ginsberg.' They call him a hero and I agree.

The Beat poet Allen Ginsberg was right to illuminate the vast discrepancies of wealth distribution in the world, the contrast between the bloated bellies of millions in New York alongside the starved bloating of stomachs in Bangladesh in the early 1970s when war and famine hit with devastating force (1996: 209). Today, these oppositions still hold to some degree, though the inequalities and privileges are weaved throughout the world in new ways. Disadvantage is sutured into the social worlds of the Global North and South, a fact that makes any simple allocation of blame or accountability difficult to sustain. Climate change is taking place in a context of historical violence. It is neither the single cause of displacement and migration, of poverty and dispossession, nor something that can be readily dismissed as irrelevant.

How can we begin to articulate the form that international responsibilities of hospitality should take in relation to this issue? It is necessary here to delve into some of the more theoretical and philosophical issues at stake before exploring them in relation to the case of Bangladesh. Dominant or analytic ethical theories concerning our responsibilities for the effects of anthropogenic climate change, as detailed in Chapters 2 and 3, tend to draw on a tradition that holds Reason and the autonomous individual as core conceptual linchpins. Out of these ideas 'distributive justice' and 'intergenerational responsibility' agendas are articulated, with specific obligations set out and rules to be followed.

The concept of responsibility that I am drawing on is always about negotiating, rather than simply abiding by, rules or guidelines. The idea that responsibility requires negotiation, rather than mere rule-based application, accords with a Derridean concept of the decision. An action is not a decision if it applies a rule without deviation over and again. The action would be considered automation, not decision-making (Derrida 1999; Mansfield 2013; Sokoloff 2005). Decisions always involve some element of difference, some variation; a challenge to the coherence of the self (being-at-home) and, with this, some gesture towards the unknown (Derrida 1999, 2000; Fagan 2013). Indeed, more radically, decision is 'of the other' even if this is the 'other in me' (Derrida 1999: 23). Decisions regarding hospitality are an example of this negotiation with the unknown:

> Hospitality, if there is such a thing, is not only an experience in the most enigmatic sense of the word, which appeals to an act and an intention beyond the thing, object, or present being, but is also an intentional experience which proceeds beyond knowledge toward the other as absolute stranger, as unknown, where I know that I know nothing of him. (Derrida 2000: 8)

What are we doing then? We are always confronted with situations that demand something in excess of the rule and reach beyond knowledge and the known. Yet at the same time, there must be some norm. Nick Mansfield rightly points out that when we rest on 'hyperbolic generosity' we offer, in the end, very little. Instead, 'we need a plan' (2012). That is, there must be categories and criteria, an 'order of priority', in Levinas's schema (1998: 156). The social world gives rise to competing rights, which then demand that we consider the unique conditions that such needs emerge within. I am alluding to the movement between the universal norm and the specific dynamics of the event in question. Going between these two poles, without rest, is what is needed to enact responsibility (as incomplete as this will be).

Norms, categories and criteria are necessary in order to recognise and organise emerging types of human movement related to climate change. Norms also enable the organisation of resources and responsibilities to provide assistance according to a hierarchy of needs. However, because the journey of each person is unique, the response to the arrival or call of the Other (or whatever form the demand takes) must be attuned to these differences. Recognising the uniqueness of the journey of the stranger invites the possibility of a response that challenges our tendency to rely on legal categories and criteria characterised by uniformity. Being forced to respond outside of strict codes of conduct is also necessary in order to qualify as a decision 'of the other'.

Moreover, flexibility and an acceptance of the risk of unknowability are needed in a context in which even our best science cannot, without some doubt, predict what the future will bring in terms of weather events and sea level rise. We cannot know in advance what will happen and as a result we must embrace the discomfort of uncertainty.

Inspired by Levinas, Derrida's notion of responsibility is Other-oriented. Fagan refers to the implications of this, including the notion that there is a 'gap' in identity (2013: 78). The 'gap' in identity is what creates the conditions for perpetual responsiveness to the Other. It is the rupture or interruption of selfhood by the Other that is the condition of dwelling (and thus of thinking and of Being).[12] The self and the sovereign are not air-tight entities, but socially constituted and responsive.

Responsibilities of Hospitality

Answering for our dwelling place involves the negotiation of responsibilities of hospitality. When we reflect on our conceptual history of hospitality, especially if we take it in its religious contexts, we tend to privilege unconditional hospitality or the sort of generosity that aims to give everything to the guest or visitor: to say, you are in need or I am at your service. This, it might be claimed, is the most ethical thing to do. It is to give before taking; to provide for rather than keep only for oneself. Levinas dedicates his entire ethical paradigm to this relationship between the self and Other. *The Law* of hospitality as unlimited is precisely what underpins Levinas's notion of an ethics of and for the Other (elaborated in Chapter 2). In *Totality and Infinity*, Levinas writes of the distinction between politics and ethics, the latter concept collapsing into a broader religiosity:

> Politics tends toward reciprocal recognition, that is, toward equality; it ensures happiness. And political law concludes and sanctions the struggle for recognition. Religion is Desire and not struggle for recognition. It is the surplus possible in a society of equals, that of glorious humility, responsibility, and sacrifice, which are the condition for equality itself. (Levinas 1969: 64)

If we substitute ethics for religion, we have a scenario in which politics and ethics find themselves in a relationship of inextricability. But this relationship is necessarily hierarchical for Levinas. Ethics precedes, makes possible and troubles politics. Ethics is transcendence; it is the overflow, the excess or infinity. Politics is totality and when closed off from ethics is totalitarian. The overflow which characterises ethics unfolds in the extension of responsibility, sacrifice and humility, without reserve. In Levinas, the encounter with the face of the Other is the pivotal event for the activation of such an ethic. Moreover, this encounter is never a choice: we are always already in a relationship with the Other as a condition of our sociality. Woven into the fabric of our social existence is the primacy of being responsible of and for the Other.

This fact is complicated by the reality of being in relation to many others. To account for this, as I discussed in Chapter 3, Levinas tells us that the encounter with the Other is also always an encounter with what he terms the 'Third' – the other Other. This figure of the Third performs the foundationally disruptive role of interrupting the purity of the face-to-face encounter. In so doing, the immediacy and necessity of ordering, discrimination and distribution come to the fore in how rights frameworks operate (Levinas 1998: 156). In other words, we are confronted with all the impurities of politics, which are, in actual fact, there from the beginning.[13] This is precisely the ambivalent binary relationship between ethics and politics.

Emerging from this philosophical backdrop, for Derrida, hospitality is the negotiation of an unconditional principle of openness and incalculability and a conditional logic which demands attention to calculation, limitation and rules. Derrida terms this the 'double law of hospitality' noting that this situation is dynamic and contingent. These structural conditions constitute the 'unstable site of strategy and decision' (Derrida 2005: 6). The promise of unconditionality is impossible but necessary if we are to follow Derrida's deconstructive logic all the way through. It can never be satisfied or fulfilled, yet it cannot be dismissed or disavowed. It must be negotiated constantly, and from this decisions must be made. Mansfield's understanding of the decision, following Derrida, is that it is always generated from the unconditional but cannot exist in this space. As such, decisions must be institutionalised or enacted in limited and risky spaces: 'Every institution must take place in relation to what undoes it, and there is nothing other than these institutions without absolute ground, which are always destined to be undone' (Mansfield 2012).

We may continually posit rules of hospitality, but these are subject to change. In other words, there is always calculation and determination in decision, but this is perpetually haunted by what exceeds it: the unknowable, the indeterminate, the incalculable (Mansfield 2012). Moreover, for Derrida, decision occurs at the threshold. Decision marks a break with what came before and a leap into a future unknown. This is the risk inherent in decision; because it breaks with a tradition, or logic of institutional modality, it refuses the safety of the programmatic response. This is why it has been argued that it is impossible to deduce a normative ethics from Levinas or Derrida (Derrida 1999: 20; Fagan 2013).

IN PRACTICE: AUSTRALIA–BANGLADESH AND THE POSSIBILITY OF WELCOME

The previous sections have detailed the philosophical resources we have available to us for rethinking responsibility and hospitality through poststructuralist thinkers such as Levinas and Derrida. I have acknowledged the difficulty we face if we try to create normative programmes from these ideas. It is now important that we consider how Derrida's work on the tension between calculation (that is, doing our best to predict and therefore organise) and the incalculable (or the inherent unknowability of the future, of what is to come, of the uniqueness of the situation) has the potential to shape or influence how we respond politically to the challenges of climate mobility. Despite the temptation to side with the excessive generosity of unconditional welcome, we must instead (and perhaps more modestly) work to 'improve the conditions of conditional hospitality' rather than attempting to attain an impossible state (Derrida 2001b).[14]

In light of these ideas, I will outline one possible policy related to the development of new 'pro-poor' economic migration schemes. In doing so, I will consider what a bilateral agreement between Australia and Bangladesh might look like. My focus on a bilateral arrangement is strategic and political, but should not be taken to exclude regional and multinational negotiations as part of the solution to flows of people in the Asia-Pacific region. Australia is a considerably privileged nation-state within the region and a high-emitter per capita of greenhouse gases, two characteristics that invoke the ethical paradigm of 'distributed responsibility'. On a pragmatic note, small-scale bilateral agreements concerned with labour migration are possible with the Australian government, as per the KANI arrangement with the government of Kiribati (discussed below). It is for these reasons that I advocate a bilateral proposal as a starting point.

In an Australian setting, scholars have begun to formulate what managed climate migration might look like. For instance, migration expert Khalid Koser has proposed the expansion of existing immigration policy to include environmental migration. He has argued that:

> What is required is a national policy framework on environmental migration with three main components: continuing support for multilateral agreements on environmental migration; capacity-building in origin and transit countries; and national legislation for environmental migrants arriving in Australia that leverages existing labour migration programs and targets a limited number of countries. (Koser 2012: 1)

Koser's suggestions encompass changes in current governance and legal migration processes. He contends that there is scope built into the existing migration programme to facilitate its expansion to include multilateral agreements with numerous other nation-states, developing labour migration streams. In other words, Koser emphasises the importance of managing migration, a process that allows the nation-state to normalise and contain the impact of human movement by integrating such peoples into economic agendas.

A positive implication of Koser's national environmental migration framework is the potential it has to redress international inequities and intergenerational injustices. This can be achieved by ensuring that the multinational agreements are signed with countries in the Asia-Pacific region, an area that will bear the brunt of climate change impacts. In this way, we could say that with some further ethical guidance, this platform may conform to the imperatives of CBDR. Moreover, if labour migration categories were to be utilised in larger climate mobility agendas, there would be a greater likelihood of promoting the rights, safety and dignity of migrating peoples.

The potential to contribute to a greater sense of equality in migration policy is significant. However, I would suggest that if we assessed this sort of proposal in accordance with the relational ethics of dwelling that I have developed in this book, there is a major limitation of Koser's (very reasonable) policy proposal. That is, the power to 'welcome' is entirely in the hands of the host state. The 'opening' towards the Other is tightly controlled and measured. Moreover, there is a danger that if we stick to contemporary migration frameworks, we reiterate middle-class migration structures. Nonetheless, we must create normative possibilities for safe passage and migration for climate migrants. Middle-class labour migration is one model. Next, I explore the possibility of pro-poor migration programmes as a more socially just alternative to middle-class migration, even though this too is controlled and regulated by the host state.

Pro-Poor Economic Migration

In Derrida's work, when the act of hospitality is institutionalised it becomes subject to the demands of a programme and can no longer be considered hospitality in the unconditional sense. Derrida makes this statement because he is determined to remind us that hospitality always contains the possibility of being surprised by the unexpected. Consequently, any programmatic response fails to understand what foundationally constitutes 'welcome' (the inability to predict and know in advance). The practice of hospitality therefore contains a paradox. One the one hand, we must predict and plan. On the other hand, we should not predict and plan, but should remain open to being surprised. If we are not confronted with the unknown, there is no hospitality. However, and equally serious, if we fail to get involved in designing hospitable structures, institutional included, we actually risk the worst kinds of violence because we turn away from the Other altogether.

The moment hospitality is part of a national policy or law it becomes a programme. As a programme it is subject to the flux of politics, and as such it is open to alteration. As Mansfield explains, institutions are perpetually exposed to that which undoes them as well as enables their existence (2012; see also Sokoloff 2005). It is this contingency and dynamism that facilitates the reform of institutional responses. Unfortunately, this condition of alterability is no guarantee that structural change will be just or compassionate, which is why we must remain vigilant in recalling the originary relation between self and Other that Levinas emphasises.

The concept of pro-poor migration emerged in interviews I conducted with NGO employees in Dhaka in July 2013. Its emphasis is on reaching the poorest members of the Bangladeshi society, who are usually so marginalised that they find themselves unable to access policy solutions. Extreme marginalisation means that the poor regularly find themselves migrating without their rights being recognised, and thus without any protections in place. Migration for the poor is difficult and dangerous leaving all, but especially women and children, vulnerable to violence. The costs of migration, relocation and resettlement are extensive and the possibility of movement outside of local regions near impossible.

The proposal for pro-poor migration attempts to undo the privilege inherent in global systems of movement. Such systems tend to promote and foster middle-class, skills-based migration into destination countries such as Australia.[15] Thus, pro-poor migration makes sense if one is interested in reforming migration policy guided by social justice imperatives. However, it is not an uncontentious proposal. Why train up citizens only to send them abroad when Bangladeshi health and education systems need the skills base? Out-migration from 'developing countries' has long been an issue that has divided the development community, which, as outlined in Chapter 1, emphasises the local as the site of belonging. Nonetheless, migration has contributed to economic transformations across the world, particularly in the form of remittances. In 2005, it was estimated that US$167 billion was sent in remittances, a figure twice what it was only five years earlier in 2000 (Skeldon 2008: 8). This total does not take account of informally channelled funds. Bangladeshi migration expert Siddiqui estimates that more than 50 per cent of remittances are informally transmitted to Bangladesh (Skeldon 2008: 8).

Implementing pro-poor migration will require extensive preparation and financial resources. It will also need to be guided by the logic of Other-oriented rights. As part of this, Australia would need to sign and ratify the *International Convention on the Protection of the Rights of All Migrant Workers and Members of Their Families* 2003. This covenant calls for the recognition of the human rights of migrants and expresses the need to recognise the specific protections migrants, particularly those 'illegally' in another country, need while in foreign territories. Ratification of this convention would be an acknowledgement of Australia's obligation to meet basic structures of welcoming. At present, Australia has definitively closed the door on the possibility of contributing to financing schemes that recognise migration as adaptation. Moreover, at the domestic level, Australia's immigration system remains committed to middle-class migration.

There are complexities that will need to be negotiated when designing pro-poor migration programmes between Australia and Bangladesh. Importantly, climate change will occur at varying timescales, which means that how displacement and migration manifest will also vary. Options for migration for the affected Bangladeshi person would need to be short- or long-term, based on the possibility of their return home. Returning home will depend on factors such as the long-term sustainability of the environment. Should the migration be short-term or even seasonal, the benefit for the migrant is that that any skills they pick up could be of use to the local community upon their return to Bangladesh. In other words, this form of migration may create the possibility for new livelihood strategies for people who in all likelihood would not be able to access migration channels otherwise. Because the positions to be taken up in the Global North would need to fill employment shortages in specific sectors, there would need to be education, language and skills development as a part of the training for those migrating.

A small-scale example of something akin to this is a project between Australia and Kiribati, funded by the Australian government and administered by AusAID, the government's international aid body. The Kiribati Australia Nursing Initiative (KANI) was a pilot project worth $20 million that ran from 2006 to 2014, with a total of 90 Kiribati participants. Three cohorts of women were trained in Australia, obtaining diplomas in nursing with opportunities for employment thereafter. An independent review of the initiative refers to it as an attempt at addressing the combined stresses of population growth, urbanisation and local unemployment in Kiribati, as well as climate change (Shaw et al. 2014: 5; see also McNamara 2015):

> KANI is a unique, bold and innovative model for 'doing development' in small, environmentally fragile Pacific island countries. The KANI concept remains strongly relevant to the sustainable development needs of Kiribati and to its adaptation policy of 'migration with dignity' to address serious climate change imperatives. (Shaw et al. 2014: 11)

Indeed, in the Kiribati policy statement concerning labour migration, this programme is highlighted as an example of its policy of Migration with Dignity required in response to 'climate change threats to livelihoods at home' (GoK 2015: 2).[16] However, the use of this language by the Australian government was not apparent in any literature reviewed by this author.[17]

The training required for Bangladeshi migrants from the poorest sectors would need to be paid for by Australia. We can view pro-poor climate migration as a compensatory measure extended as a result of our contribution to anthropogenic climate change. Further, we can see it as a means of social development in Bangladesh, assisting with skill acquisition and economic opportunity (through remittances), though as Shaw et al. note in their review of KANI, such long-term benefits are not readily apparent and will only be recognisable after some years (2014: 21). Even as it is caught up in the power relations that shape global flows of migration, pro-poor migration can be applied in a way that is ethically responsive to the emerging condition of widespread and uneven climate crisis mobility. It aspires, perhaps, to what Derrida refers to as a lesser violence (1978; see also Sokoloff 2005; Hägglund 2008).

This is, of course, a conditional and problematic hospitality. In practice, pro-poor migration may look like a lucky dip or lotto (as the US Green Card system is called). The fact that there would be limitations on numbers does create a condition that risks drawing arbitrary lines between those who are entitled to migrate and those who are not. In the case of the KANI initiative, the women chosen to be part of the programme were already proficient in English, reflecting their relative privilege in Kiribati society (Shaw et al. 2014). We might then undo any attempt at justice that we are trying to promote (at least at the level of the individual experience of exclusion or rejection). Not all victims can be accommodated and processes of selection would be both exclusionary and open to the risk of corruption.

In light of this, we can see a number of serious limits. Firstly, pro-poor migration sits uneasily next to contemporary global migration trends and will be difficult to generate support for. Much managed economic migration to the Global North is targeted and represents a very conditional mode of 'welcome' which overwhelmingly benefits the host country. Secondly, selection processes contain within them an inherent violence in the work of including and excluding. This is a condition of classification of which we need to be constantly aware. So even in striving to be ethical we are bound up in potentially violent processes. Last, but not least, we exist within the capitalist system for which we need to be responsible for the damage to the climate it is implicated in.

Yet at the same time, pro-poor migration subverts and reiterates global relations of power and systems of dominance. We cannot know in advance the implications of the policies we develop or the manner in which they will be taken up and reshaped according to social and cultural forces. The conditions of hospitality can always be critiqued because they manifest through power structures. Moreover, it is the fact that we 'cannot know whether our aspirations are just' that we are compelled to continually critique, destabilise and reformulate (Sokoloff 2005: 344). My interest is in proposing ethically grounded forms of social and political transformation, despite the practical challenges that invariably arise. In sum, then, if organised well and in accordance with applicable rights regimes, managed pro-poor migration offers safe passage for peoples across international borders, as well as economic opportunity, and – if written into the agreement – assistance with resettlement. Pro-poor labour migration is conservative and radical, planned and open to alteration, embedded in forms of exploitation and an attempt at a socially just migration policy.

CONCLUSION: COMMON GROUNDS

Climate change presents the global community with challenges that will need co-operation, generosity, responsibility-sharing and sacrifice. This chapter has paid particular attention to the case of Bangladesh, where local adaptation and migration as adaptation are complex and urgent topics. In the early sections of the chapter, I outlined the *in situ* adaptation challenges the country faces. I emphasised housing, land and property rights and the pivotal role each plays in ensuring 'adaptation with dignity'. Security of land tenure in rural regions lessens the likelihood that people will need to migrate. Meanwhile, some assurance that one's home is secure and safe in urban contexts provides the foundations for political, economic and social engagement.

There is a great deal of work going on, at both the grassroots and the international governance levels. I have not been able to encapsulate all of this here. And there are large gaps in this story. Indeed, in reality, the case of Bangladesh deserves a whole book. However, what I have highlighted is the need to fully recognise and integrate human mobility into our adaptation platforms. This takes shape distinctly in domestic and international settings. In Bangladesh, development programmes must begin to offer support for peoples on the move or chronically displaced. This means that the international community must commit to financing internal migration schemes. Such schemes would support the facilitation of orderly and safe migration and internal relocation. Early practical efforts can be seen in the *Cancun Agreements* and the draft report recommendations of the human mobility expert panel, discussed in Chapter 3. Social support for migrants in Bangladesh is already happening in a patchwork manner, but it must be operationalised as a norm in order to garner economic contributions from the Global North under the ethical obligations inherent in the CBDR framework.

Local adaptation is only one (very complex) thread of the story of climate impacts for Bangladesh. I have endeavoured to articulate an ethics of hospitality regarding international or cross-border climate mobility in the latter sections of this chapter. I have drawn out the case study of Bangladesh by focusing on how Australia could foster a relationship to the social fact of mobility that takes account of, and indeed privileges, efforts to relieve the suffering of the Other. Even as I have put forward a managed migration model, we must keep in mind that human mobility is both something that can be calculated and planned for, as well as a social practice that is inherently risky, incalculable and unpredictable. We cannot always control mobility! Consequently, we cannot rest solely on predetermined migration streams, and must be responsive and responsible to the one on the move, whether their movement is deemed legal or not.

Chapter 5

IN AND OUT OF PLACE: THE CASE OF THE TORRES STRAIT ISLANDS, AUSTRALIA

'The future belongs to ghosts.'

Jacques Derrida[1]

'But this is our birthplace. We born here, we live here, we die here. Only when high tide wash out this island, that's the only time to move from here.'

Saibai elder[2]

HAUNTED HOUSE

The *future* belongs to ghosts.
The future *belongs* to ghosts.
The future belongs to *ghosts*.

Derrida's spectral logic is confounding. Ghosts haunt the present. We are told this. Ghosts from the colonial past impose themselves into the 'now'. They disrupt any sense of complacency that may be prevailing. Ghosts wreck the promise of calm. They hinder any efforts to expel the past from the present. But does this mean that the future is the possession of ghosts too? Wait, does this mean that the present is possessed by ghosts? Does Derrida's suggestion, that the future belongs to ghosts, suggest that all acts of possession are haunted? Does this haunting render 'our' possession unstable or insecure; foundationally interrupted? Is the haunting there to hold us to account for the past and looking forward, to the future?

Ghosts persist when the dead are not able to rest in peace. They are forceful when a death is unjust, premature, avoidable, undignified. A good death does not, colloquially speaking, remain in the form of a ghostly presence. Sadly, I fear that the ghosts of the future will multiply. There is no proper resting place permitted by the neo-colonial state for such transient figures. But can ghosts also persist as a reminder? Can ghosts be a trace of the force of responsibility for maintaining a sense of commitment or respect for the past, and for maintaining culture into the future? Can these unsettled, unsettling figures be a creative force urging transformation? An intergenerational justice, perhaps?

I have been arguing that adaptation is a political discourse. Indeed, adaptation is caught up in the pre-existing power relations, norms and beliefs that a society holds. The impacts of climate destruction in the Torres Strait Islands (TSI) return us to questions that have haunted human beings without rest: how should 'we' live? What is a 'worthy' life? And what sort of future do 'we' want to build? In an Australian setting the answers to these questions are affected by our response to two further questions stained by our history of colonial violence: whose future counts? Who are the 'we' who matter? If non-Indigenous Australia cannot accept the reality of climate change, we will be implicated in future injustices towards Torres Strait Islanders (TSI) amongst others in our community.[3]

In this chapter, I provide an overview of the impacts of climate change in the Torres Strait. Within our dualistic model of political organisation, as Australian citizens, TSI peoples have the right to expect that national resources are, and will continue to be, committed to building a sustainable and flourishing environment for its people. Moreover, as Indigenous peoples, Islanders are entitled to the specific rights contained in international covenants, including the *United Nations Declaration on the Rights of Indigenous Peoples* 2007. In this respect, they have every right to demand that an 'adequate', and locally generated, adaptation plan be supported by the state of Queensland as well as the Australian government. The focus of adaptation work in the TSI is local and aimed at retaining social and cultural connections to land, sea and community.

Adaptation must focus on the preservation of the local. Nonetheless, Islander peoples have a history of migration which informs the sorts of mobility practices taken up by locals today. Migration between the islands and over to the mainland has a long history in the region, complicated by the experience and impacts of colonialism. To a large extent, migration is an everyday practice. This reality does not invite apathy or negligence concerning the imperative to support *in situ* adaptation across the islands. That is, the reality of a dispersed TSI population does not negate or minimise the importance of the relationship between land rights, identity and belonging for all Islanders, regardless of where they end up living.

Following the discussion of local adaptation and the ordinariness of migration, I will turn my attention to the topic of communal relocation. The impacts of climate change are likely to render a number of islands, particularly Saibai and Boigu, uninhabitable within the next century. Should this prediction come to fruition, cultural loss and trauma experienced through colonisation will again be experienced by Islanders. What should Australia do to prepare for this possibility? Should compensation be available (and what would it look like)? We should not just assume that dispossessed Islanders will desire relocation to another island or the Australian mainland. In fact, the formal policy of the Torres Strait Regional Authority (TSRA) is relocation only as an absolutely 'last resort'.

In this instance, I engage with an Other-oriented ethics to argue that the Australian government should adopt a stronger policy for Aboriginal and Torres Strait Islander political representation and autonomy. How might autonomy or some form of national political representation affect local adaptation planning, funding and implementation? The importance of autonomy may be even more pronounced in the future, particularly if the spectre of communal relocation emerges. In an era in which the lives, lands and cultures of TSI are threatened once more, we must foster policies that recognise and draw on local knowledge, leadership and resilience in the work of adaptation. This is our responsibility towards the future; our intergenerational responsibility. The ghosts demand this of us.

As a non-Indigenous Australian writing about this aspect of climate change, I am cognisant of the power relations in play. In particular, I am aware that my social and economic privilege affords the opportunity to write and be listened to. From the outset, then, I want to emphasise that relocation is considered a 'last resort' for Islander peoples. Moreover, it is vital that any articulation of a policy proposal on this issue arises out of substantial and ongoing community consultation. My aim in contributing to this discussion is, firstly, to unpack the political and historical context that these issues are unfolding in and against. Secondly, from a non-Indigenous perspective, I want to promote an ethical dialogue that is responsive to, and takes responsibility for, the ghosts of the colonial past. As per the framework I have been developing, this necessitates a series of cultural, affective and social changes at a personal level, not merely higher up in government. Climate change is an intergenerational issue and is often understood in this manner. Usually, this responsibility is cast towards the future. However, the past haunts too. These unsettled and unsettling figures demand our attention, and call upon us to redress the harms of our history. To close the chapter, I deploy Derrida's concept of 'hauntology' to remind Australians, and all peoples, of these obligations towards the Other, as non-deferrable and unrelenting.

CLIMATE IMPACTS IN THE TORRES STRAIT ISLANDS

Worldwide, Indigenous peoples contribute the least to human-induced climate change, yet are among the most vulnerable to its effects (United Nations Permanent Forum on Indigenous Peoples n.d.). Across Australia, Indigenous peoples constitute about 2.4 per cent of the total population (IPCC 2007). Over 100,000 of these people live in remote regions (Green et al. 2009: 3). Relative to the Australian mainland, the TSI is populated by only a small number of people. There are approximately 7,000 TSI living in the region, spread across sixteen of the islands (Australian Human Rights Commission 2008: 229). According to Australian Census data released in 2011, of the islands exposed to the effects of climate change, Boigu has 283 residents, Saibai 337, Warraber 247, Iama 311, Poruma 166 and Masig 300 (TSRA n.d.). The climate impacts set to affect the region include rising temperatures with an increase of between 0.5 and 1.5 degrees Celsius anticipated by 2030. According to the IPCC, this may jump up to between 1.5 and 3 degrees Celsius by 2070 (2013; Green et al. 2009: 19). In conjunction with this rise in temperatures, changes in the rainfall patterns will likely result in longer wet seasons and less predictable rains during the dry seasons. Coastal erosion and flooding will also affect the islands, causing damage to homes and main infrastructure. In addition, strong storm surges and high tides risk washing out community structures, such as homes, as well as culturally significant sites, including local graveyards (Green et al. 2009: 25).

In the TSI it is the predicted rise in sea levels that will cause the most severe forms of physical, psychological and cultural damage. In the long term, the IPCC has indicated that a high-end prediction of sea level rise is 82 cm by 2100 (IPCC 2013). Two islands close to Papua New Guinea, Saibai and Boigu, are at risk of being completely uninhabitable by 2100. When I visited Hammond Island, an inner island close to Thursday Island (the latter is the administrative hub and major population centre of the region), I saw the way that the local graveyard was close to the shoreline. Locals expressed their concern to me over the graveyards and hypothesised what may become of their ancestors. For some, the impact of climate change will threaten their land, livelihood and culture.

Indigenous peoples and communities have been extremely resilient in the face of the violence that has marked, and continues to mark, Australian society's relationship towards them. But resilience is not an infinite or unending resource. When conditions fail to ease, or even increase in difficulty, the individual and collective quality of resilience wears thin. Resilience must be socially cultivated, supported and maintained. This is a core component of what our adaptation planning should be enabling. The climate change challenges that Indigenous Australia faces are disproportionate to those that non-Aboriginal Australians are facing. This is due to a range of social, political and environmental factors. Pre-existing social disadvantage in remote locations affects the health, education and general support services available to many Indigenous peoples. Remote communities also face increased temperatures, increased risk of disease (for example, mosquito-borne disease), extreme rainfall and flooding, coastal erosion, sea level rise, and damage to local infrastructure. This position of disadvantage is doubly the case in relation to climate change. Many remote communities live in deserts or on islands. Both of these landscapes are inherently environmentally exposed. The colonial history of Australia, marked as it is with the legacy of dispossession, further compounds the situation of exposure.

In 2007, a popular Australian Labor Rudd government introduced the Department of Climate Change and Energy Efficiency. This department worked hand in hand with researchers, communities and policymakers and produced a flurry of research and reports on the impacts of climate change. A number of these documents detailed the implications for the northern regions of Australia (in particular, see McNamara et al. 2012; Green et al. 2009). Recommendations were implemented in the TSI, with researchers from James Cook University in Queensland leading an adaptation programme across numerous islands, particularly Erub and Boigu (McNamara et al. 2012). During this time, government funding for the construction of major infrastructure like sea walls was promised and withdrawn, tangled up in complex governance structures and subject to bureaucratic bickering between the Queensland state government and the federal government. Inter-governmental cooperation has been noted as vital to addressing climate change, with the Australian Human Rights Commission arguing that the TSI region is the 'inadvertent litmus test for how the Australian and Queensland governments distribute the costs and burden of climate change' (2008: 232).[4]

LOCAL ADAPTATION, LOCAL RESISTANCE: ROOTS AND RIGHTS

The Australian Human Rights Commission has provided an extensive study of the rights implications of climate change for the people of the TSI. Its report determines that several significant rights are at stake, particularly rights to life, water, food, health, culture, as well as to a healthy environment (2008). In turn, the report highlights the importance of the *United Nations Declaration on the Rights of Indigenous Peoples* 2007 in responding to climate change, noting its emphasis on the 'full participation and engagement of Indigenous peoples in the development and implementation of national and international policy'. This theme of empowerment through political participation (in conjunction with grassroots adaptation plans) will be highlighted in due course.

As I explored in Chapter 1, the UNFCCC advances an understanding of adaptation which emphasises practical or pragmatic measures. However, psychological, emotional, and intangible cultural and spiritual implications are more difficult to contain within this formulation. These seemingly less material things are, nonetheless, advocated for within existing rights frameworks. In addition to the medical and infrastructural problems I have described above, the myriad psychological and emotional impacts of climate change are beginning to gain some recognition as valid. Slowly, we are coming to appreciate that the relationship human beings have with the earth goes beyond cost–benefit economic modelling and consumerist intent.

Attachment to place is bound up with identity and belonging. Indeed, attachment to land implicates issues of safety and wellbeing (Gelbspan and Thea 2013). The relationship between land and identity is especially pertinent in an Indigenous context. In Australia, colonial power attempted to sever this connection for Aboriginal and Torres Strait Islanders. The violence of colonialisation was predicated on the individual and communal dispossession and dislocation of Indigenous peoples from home/lands. Thus, we must be particularly sensitive to the importance of land when we begin to articulate our national responses to climate change. McNamara et al. argue that we must recognise the centrality of land to identity when they write that 'connections between the land, sea, environment and culture are paramount to identity, livelihood, and sustainability . . . environmental changes associated with climate change projections potentially compromise this core cultural identity' (2012: 7). A right to health requires a right to the preservation of culture and thus of the earth itself. That is, embodied health is culturally embedded in the health of the land and environment more generally.

It is important to understand that the physical or geographical changes occurring in a place like the TSI have implications which exceed the tangible damage identifiable in eroded coastlines and flooded townships, or in normative markers of 'development' such as lifespan measured in years. These things are, of course, enormously important, but cannot quite capture the intangible cultural losses generated when land is dramatically affected by climate. Indeed, the material damages expose a deeper and more profound impact for Islander communities. An oft-cited example is the flooding and feared washing away of culturally significant sites on small islands, such as graveyards. Culture is always already in a process of transformation. Like all things it is impermanent and subject to the relentless changes that accompany the passage of time. These changes are responsive to environmental, political, social and economic circumstances. This process of 'organic' change must be distinguished from mandating the imperative to 'adapt' one's culture and belief systems in response to climatic changes. Indeed, the line between adapting and enduring loss is ambiguous.

This is not to suggest that TSI peoples are not exercising any autonomy or expressing their resilience at present. Local communities at the grassroots level are planning for the future in ways that recognise the changes occurring to land mass, weather systems, breeding patterns and coral reefs. This planning is aimed at preserving their ancestral lands and cultural roots. For instance, conversations that I had with councillors emphasised the work that the local councils are trying to do with planning infrastructural works away from the vulnerable coastlines.[5] This work is being done in order to ensure the sustainability of future generations. Though, as is too frequently the case, limited finances were raised as an obstacle to this ambition. Even though local control over education and health services has been attained, the region remains economically dependent on the state and federal governments. Financial support for the region is entangled in various government bodies. It is for this reason that, as the Australian Human Right Commission has pointed out, the area provides the 'litmus test' for how levels of government (and departments within these relevant levels) will work co-operatively for climate justice outcomes.

During the colonial era, state governments had supreme authority over 'Aboriginal Affairs'. The Torres Strait was considered a part of Queensland. As such it continues to be enmeshed in its bureaucracy. As a result of this historical context, the Torres Strait Island Regional Council (TSIRC, which is the umbrella organisation for the individual island councils), is subject to the local government laws of the state of Queensland. This body also receives both financial and administrative support from the state of Queensland. Indeed, 'control over resources and land, as well as future development of the Torres Strait region, lie primarily with the State [of QLD]' (Babbage 1990: 314). At the federal level, the TSRA was formed 1994. In the 2014 Federal Budget, over a four-year period $3.5 million was cut from the regional authority's budget (TSRA 2014).

The development of today's complex bureaucracy is the outcome of a combination of colonial administrative power and local TSI resistance to these structures of power. The history of Islander agitation for autonomy goes back to the 1930s when a major strike by workers in the fisheries sector resulted in changes to governance that were 'without parallel in colonial Australia' (Beckett 1987: 54). As Osborne writes, 'the Torres Strait Islanders had nothing apart from their labour with which to demonstrate their refusal to accept a cultural death under the outsiders' law' (2009: 31). An outcome of these strikes included the development of a consultative process in which Islanders elected local representatives who were then invested with some 'local autonomy' under the *Torres Strait Islander Act 1939* (Beckett 1987: 54). Appeals to local roots and the primacy of cultural belonging came to support calls for political autonomy and the extension of decision-making powers.

In other words, during the colonial era, external authority was fiercely contested in the region. Autonomy, self-determination and resilience were evident from the time the area was subjected to colonial control. The use of a cultural roots discourse served, and continues to serve, as a critique of the dominant national relations of racialised power. This in turn promotes local empowerment. Indeed, it came to take hold as a sort of anti-colonial struggle in the face of cruel and degrading impositions placed upon the local people by the colonisers.

The movement of Islanders was tightly regulated by the colonial administrations in the nineteenth and twentieth centuries. For instance, during the late 1920s and into the 1930s, Islanders had to seek permission from the white authority (usually the teacher on the Island) to travel to Thursday Island or to other islands in the region (Osborne 2009: 27; Beckett 1987: 58). Travel to the mainland of Australia was strictly prohibited (Osborne 2009: 27). Osborne articulates the racialised logic underpinning this period which forms an inherent part of paternalistic governance:

> For three decades, legislators with a 'mania' for paternalistic control ... extended the Protector's powers and frustrated the people's aspirations for self-regulation. Increasingly white people administered island affairs with a racist mentality of their own supremacy ..., and with no respect for the culture of those they administered. (Osborne 2009: 28)

The obsessive and unrelenting control asserted by the colonial government extended from the regulation of movement and money, to the punishment of Islander men who appeared to be engaged in relations with white women (Osborne 2009: 28). Before this, mobility between islands had its own cultural dimensions. People came and went; trading with neighbouring Islanders continued, adhering to local customs and practices.

Against this backdrop, local resistance took on a collective form. Despite cultural diversity between the islands, the strategic use of a unified and enrooted identity was solidified. This manifested in the identity category of Torres Strait Islander during the aforementioned political activities of the 1930s. In other words, this identity category emerged in the twentieth century as a response to the colonial administration's attempts to control Islander identity, shatter any sense of solidarity, and culturally assimilate Islander people into white Australian society through the education system.

The long struggle that many have endured to obtain land rights and more recently sea rights in the Torres Strait is further evidence of the remarkable resilience of Islander peoples. Collectively, the communities have fought for and received recognition of native title and sea rights. The connection to the land and sea as co-constitutive of one's individual and communal identity also calls for an understanding of dwelling as related to the care and preservation of place. Funding for local adaptation in the region must be informed by the protection of land and sea rights and support for cultural roots.

MIGRATION AS ADAPTATION

While there has been a strong call for the recognition of the relationship between identity and land, Torres Strait Islanders have long been 'adapting' to the impacts of colonialism, with 'migration as adaptation' a strategy used in response to economic, social and ecological changes since the end of WWII, and particularly since the 1970s. That is, while there is a strong emphasis on local adaptation, there is simultaneously a clear enthusiasm for taking up the opportunities available via migration. In this sense, like in Bangladesh, speaking of climate migration in this context makes little sense without understanding contributing and perhaps even more significant factors related especially to economy, family networks and other opportunities. Migration as adaptation should be viewed firstly as a description of what is already occurring. Secondly, migration as adaptation is a strategy employed by people to negotiate the conditions of their lives, a historical norm (Nail 2015). That is, it should be seen as a livelihood technique with variable effects. However, in the case of future instances of forced climate-driven migration from one island to another, or from an island to the mainland, policies will need to be developed to assist with the material and psychological dimensions of this movement. It will be the responsibility of the Queensland and Australian governments to organise this in advance of, as well as in response to, climate change events.

On Thursday Island, I hear stories of migration. They reflect what is by now a cliché in poststructuralist circles: that the world is weaved together by difference. Perhaps we yearn for linearity and coherence in our narratives, whether social or psychological, local or national. Linearity promises a sense of integration and satisfaction. Yet these stories reflect the reality that each present moment is displaced by the past and future. We are haunted. Temporal dislocation is entrenched in settler colonial societies. It reaches into our national, communal and individual psyche. The impacts of these dislocations range and are deeply embedded in specific social, political and economic contexts. Up on Thursday Island, I am told of the young men who go 'south', to the 'mainland', and find themselves without an anchor. Some end up in gaol, especially those who suffer some sort of psychological internal exile. This exile is the ramification of historical dispossession or dislocation and it rages within. It expresses itself as an inability to stand still for long enough to touch the earth. I am told that they aren't bad kids, and I believe this. But their home has been stolen from them. Others take root on the 'mainland', getting jobs up and down the coast from Brisbane to Cairns. They have families, and get on with life. Many travel up to the TSI to visit family.

Not all stories of migration are despairing. After all, 80 per cent of Torres Strait Islanders live 'down south' on the mainland of Australia (Arthur 2001: 12). In the 1960s, anthropologist Jeremy Beckett estimated that only a small number, perhaps 500, Islanders lived on the mainland. It was from the 1970s onward that longer-term migration gained popularity as employment decreased in the TSI (1987: 72). In one of the few focused discussions of the migration of the region, Jeremy Hodes identified three waves of out-migration in the region. Prior to WWII, there was small-scale migration, mostly to Cairns. Once in Cairns, Islanders were removed from the general population. They were cast to the outskirts of the community and formed a small shanty called Malay Town (Hodes 1998: 3). In the period of the early 1900s, the Torres Strait Islander figure, Douglas Pitt, was widely celebrated as an entrepreneur, explorer and extraordinary swimmer, but even he found himself shifted to Malay Town when settling in Cairns (Hodes 1998).

During WWII, with the threat of Japanese invasion, Islanders were evacuated by the colonial authorities. This is considered the second wave of migration. The third wave is the biggest and most significant. It started in the 1950s and was largely economically driven (Hodes 1998: 3). The second wave does not qualify conventionally as 'migration'. Instead, it is an example of forced relocation. However, the ramifications of the war, with the inclusion of Islander men as soldiers and the movement of peoples across to the mainland, did lead to the decision by some to live away from the islands.

A local adaptation technique in the Torres Strait has long included transient, temporary migrations 'south' or even to the 'west' for employment and education. For example, journalist Chris Hammer, in his book *The Coast*, speaks with a local resident of Yam Island, Patrick, who travels to Perth regularly for work. Patrick works in a mine in Karratha, Western Australia, for a month or two at a time, returning to Yam for a week or so in between trips. His children live in Cairns where they attend high school (Hammer 2012: 48). Due to the remote nature of many of the islands, children are often sent to Thursday Island or even the Australian mainland for long periods. This has cultural implications, with some Saibai people, for instance, expressing 'worry that their cultural heritage – e.g. songs, dances and stories – are not being transmitted to the young, due to modern influences such as television and remote schooling' (Green et al. 2009: 121–2).

If we remain in conventional migration studies discourses, the main drivers of migration in the region are economic and social, rather than climate-based. The implications of these migrations remain complex and cannot be resolutely determined to be 'good' or 'bad'. Recognising the ordinariness of migrations 'down south' highlights the resourcefulness and resilience of Torres Strait Islanders. Moreover, the Australian citizenship held by Islanders ensures legal passage and free movement within the territorial borders of Australia. In this sense, the cultural practice of migration does not induce concerns regarding borders and national identity in quite the same way as the possibility of migrants coming over from the Pacific Islands or further afar does for the Australian government. However, future forced migration will require government support and funding to organise.

BEYOND ADAPTATION: RELOCATION IN THE CONTEXT OF NEO-COLONIALISM

It is not migration that is the looming issue for the region. By the end of this century there is the threat of communal relocation for the communities of the Saibai and Boigu Islands. Taking into account the history of forced removal and relocation of Indigenous peoples during the colonial regime, it is vital that we understand that this issue is sensitive and difficult. In Chapter 3, I outlined the recommendations put forward by the Advisory Group on Climate Change and Human Mobility. As part of a three-pronged approach to the diverse mobility scenarios that climate change triggers, the Advisory Group highlighted the need to develop protocols for 'participatory and dignified relocation' (2015: 3). This, of course, is a 'last resort' and should not replace mitigation and local adaptation programmes. In other words, the process of relocation requires community consultation and participation, long-term planning and significant financial commitment. The case of the Newtok peoples in Alaska, discussed in Chapter 2, indicates that this planning and financing needs to start well before any physical relocation takes place.

The *Guardian* reporter, Susan Goldenberg (2013), conveys the psychological ramifications of climate change in Newtok. She refers to the anxiety triggered by an uncertain future. Globally, it is this uncertainty regarding the future, as well as the fear of future cultural loss, that is the backdrop of existence for marginal peoples. This reference to anxiety must be approached as distinct from Heidegger's concerns, which I presented in Chapter 2. Heidegger's central thesis posits that anxiety over the future (being-toward-death) is the ontological condition of Dasein that we must come to terms with in order to dwell authentically. This is not the same as the anxiety vulnerable peoples are experiencing in the face of climate crises.

The anxiety that many climate victims potentially endure is the cumulative effect of historical and political forces that have undermined and even destroyed individual and communal futures. Structural conditions within a society are implicated in the possibility of what we might call a 'safe future'. What this means is that when appropriate institutional arrangements are in place and responsive, the likelihood of cultural, social and personal anxiety decreases. Alternatively, when institutions fail to extend support and care, this may actually exacerbate and even produce psychological distress. This is a markedly different approach to anxiety than that posited by Heidegger, who regarded it as an existential condition to be confronted and overcome.

The relationship between land, dignity and the future is at stake in scenarios involving communal relocation. For the community in Newtok, Alaska, for example, it is the lack of assurance that a safe future exists for their community, and the next generation, that has underpinned the decision to find a new home. This decisive act is not without its complications. Even if the relocation efforts qualify as 'dignified' and 'voluntary', they are not necessarily examples of untainted agency, or simple instances of empowerment by disadvantaged peoples. They may contain both of these qualities, empowerment and agency, but to focus only on these is to neglect the tremendous suffering experienced by the community members leaving their home behind.

Communal relocations associated with climate change are decisions made in the midst of crisis. You may recall that the Greek origins of this term crisis/*krisis* refer to 'decision' or a 'turning point'. Remembering that these decisions are the outcome of a crisis allows us to reflect on the role of both context and power. Rather than being the outcome of an intentional process of reasoning, decisions are integrally associated linguistically with crisis, or that which highlights the withdrawal of reason and order, of stability and grounding. When the earth beneath one's feet is literally disappearing, re-grounding is sought after.

By opening up a discussion of loss (past, present and future) I do not intend to promote a defeatist or anti-local adaptation discourse. Nor am I trying to negate the import of continued political pressure to mitigate. There are effective and empowering forms of local adaptation taking place, including in the TSI. However, climate science and its predictions of a substantive sea level rise compel us to think about loss in ways that advocate for care, justice and ethics, rather than ignorance, exclusion and denial. Voluntary and participatory relocation involves the negotiation of profound tensions between loss and transformation, and despair and hope. There is the imperative to think of the future as sustainable in new lands, and thus possible in a transformed manner. However, in so doing, there is the need to recognise the inevitability of some experience of cultural loss. This is particularly so if the move occurs without extensive consultation and local planning, or involves the loss of territory to rising waters or erosion.

Relocation in an Australian context is set against the past, and performed in 'the colonial present' (Gregory 2004). During the colonial administration the relocation of Indigenous peoples, from traditional lands to 'reserves', was driven by an assimilationist agenda. The effect was to destroy a people, to shatter the culture, and to relegate Indigenous peoples to the fringes, or indeed, to the pages of history. Indigenous peoples were systematically denied the promise of a future. Separating peoples from their homelands was pivotal to the colonial agendas. This is because, as Gelbspan and Thea emphasise, the 'struggle for land [is] an integral part of a life of dignity and a better, shared future' (2013: 14). When this relationship is broken, denied or refused, there is an experience of loss which endures long after the event of dispossession or removal. Consequently, when dispossession or profound dislocation is the historical backdrop for a community or people, the experience of loss marks each moment without relief. As Donna Green and her colleagues have highlighted:

> History has shown that relocation of Aboriginal people from traditional homelands to the larger regional centres has had a negative impact on spiritual and social health, as well as reduced opportunities to participate in important cultural activities . . . Any further centralisation of Indigenous settlements will likely continue that pattern of social dislocation and the full costs and benefits of this course of action should be taken into account in any climate change related policy development. (Green et al. 2009: 9)

From Saibai to Bamaga

Given the abundance of examples of violent and oppressive colonial policies, it is very rare to find any instances of something that does not simply reinforce this colonial history of white supremacy. Yet there is a unique story of relocation during the mid-1900s that remains difficult to integrate, if only because it does seem to suggest that the colonial administration of Queensland was able to work co-operatively with the TSI peoples. This story is the participatory and organised relocation of a portion of the Saibai people's community following WWII.

Saibai Island sits a short distance from Papua New Guinea. At only 22 km in length and 7 km wide, at its highest point it is only 2.7 m above sea level, with a flat and swampy geography. A few thousand years ago, alluvium deposits from PNG formed the small island, and sometime after this, most likely thousands of years ago, people of Papuan Melanesian descent settled the island (Babbage 1990: 6). The main town centre is subject to constant flooding in the wet season. King tides are increasingly causing disruption to households. However, there is limited opportunity for relocation to another part of the island (Green et al. 2009: 121–2). Locals speak of using dinghies in the living spaces of their homes during the worst periods.[6]

It was the dynamism of this fertile environment that led to the decision by some Saibai residents to move to the mainland in the years following WWII. These communities are now established at Bamaga and Seisia, on the Cape York Peninsula in far Northern Queensland. During WWII, many young Islander men were recruited to serve in the Australian Armed Forces. This experience exposed the men to the mainland of Australia. This sort of travel had been prohibited by the colonial administration in the decades preceding the war. Once on the mainland, these men identified Mutee Heads as a potential site to which their community could move (Poi Poi et al. 2000: 6). Very little has been written about this process. I draw heavily on the Bamaga Island Council Centenary of Federation publication on the founding of the new community, *Saibai to Bamaga: The Migration from Saibai to Bamaga on the Cape York Peninsula*, to retell this story.

At the end of WWII, the men serving in the Australian Army returned to their respective islands, Saibai, Duan and Boigu, and spoke with local elder, Chief Bamaga Ginau, who agreed that the long-term survival of his people depended on 'communal migration' (Poi Poi et al. 2000: 7). Permission to move had to be approved by the colonial government of Queensland. The community received permission to move to Cowal Creek Aboriginal Reserve, Mutee Heads, where they took shelter in the existing Army huts and utilised the local infrastructure, including a school and a church (2000: 21). In 1947, initially only a small number of the community, seven families, made the move, with most people disagreeing that migration was required until flooding 'made the water supplies brackish' (2000: 16). According to Mr C. O'Leary, Director of Native Affairs, Department of Health and Home Affairs, by 30 June 1949,

> Well over 340 Saibai Islanders are now established on the Cowal Creek Aboriginal Reserve at Mutee Heads, in temporary quarters, and the intention is, as water and buildings are provided, to remove these people as the nucleus of the population to take over these Peninsula areas. (In Poi Poi et al. 2000: 41)

From Cowal Creek, representatives of the Bamaga people, the Department of Native Affairs, the three island group representatives, and the Church of England were all involved in negotiating the decision upon the final settlement at Bamaga. According to the Bamaga Island Council, 'Chief Bamaga told the Minister that he wanted the settlement built between the two creeks in the area to provide plenty of water for his people' (Poi Poi et al. 2000: 26). On 24 July 1948, the Queensland Government, under the powers vested in *The Lands Act 1910* (operational from 1910 to 1948), reserved an area of 44,500 acres for the new community to settle. The relocation took place between 1947 and 1954. Despite its success, there were many Saibai people who decided to remain on their home island.

Because relocation remains a 'last resort' for communities, there is little open discussion of what it may look like. Indeed, there appears to be a fear that engagement with this instance of past relocation may expose the community to the charge that they should simply join the rest of their people on the mainland, as if the decision of a previous generation binds them. I suspect that it is due to this that references to this unique event are often brief and descriptive. Donna Green's inclusion of it in her extensive government report says the following:

> In 1947, the community experienced a particularly high king tide which flooded the village and destroyed many gardens. That same year, a number of families decided to leave Saibai to establish a new life on Cape York Peninsula, settling first at Mutee Heads and soon after establishing the township of Bamaga. To date this has been the only relocation of its kind in Torres Strait. Many other families, however, decided to stay on at Saibai. (Green et al. 2009: 124)

My aim in presenting a greater level of detail and analysis of the historical context is not to resign any policy-oriented solutions to following the suggestion that current Saibai people should join the established communities on the mainland. In fact, community consultation would be a priority in this case and may reveal that other options are preferable if and when climate-driven relocation is required.

How are we to negotiate the complexity of this history? *Saibai to Bamaga* is a celebratory text, positioning the ancestors of the current generation as strong, decisive and future-oriented. They had to negotiate an oppressive regime and were visionary in their thinking of a future sustainable community. According to the Bamaga Council's representation of its history, the relocation from Saibai to Bamaga was arranged by local organisations and driven by the community participants. However, the co-operation between organisations evident in this example was ultimately at the whim of the colonial administration. Had the Chief Protector changed his mind, the efforts of the local community would have mattered little.

What recourse is there in the international arena for Torres Strait

Islanders facing the worst effects of climate change? To break this question down further, in the event of cultural loss, what relevance do rights have in providing a basis for adaptive action, and even compensatory redress? The *United Nations Declaration on the Rights of Indigenous Peoples* 2007 is a non-binding text which the Australian government formally 'endorsed' in 2009, having first voted against the Declaration in 2007. This endorsement does not equate with full recognition and implementation of the rights encoded into Australia's domestic laws and policies. Like all international doctrines, the primacy of the sovereign state is clear. However, as a declaration setting out aspirational norms, several articles are particularly important for our discussion of climate change adaptation and communal relocation. Article 8(2) provides two foundational components for action. These are prevention and redress:

> States shall provide effective mechanisms for prevention of, and redress for:
> (a) Any action which has the aim or effect of depriving them [Indigenous peoples] of their integrity as distinct peoples, or of their cultural values or ethnic identities;
> (b) Any action which has the aim or effect of dispossessing them of their lands, territories or resources;
> (c) Any form of forced population transfer which has the aim or effect of violating or undermining any of their rights.

Prevention is obviously best offered through mitigation. Yet as we continue to prioritise economic 'growth', we neglect our responsibilities towards the earth and contribute further to its destruction. These destructive actions erode the capacity of mitigation of, and even sustainably adapting to, the impacts of climate change.

Legal scholar Owen Cordes-Holland has highlighted the obligation that Australia has under the *Kyoto Protocol* to reduce carbon emissions, and the manner in which the government's efforts have been inadequate. This opens the Australian government to the charge that its failure to adopt effective targets amounts to a 'violation of Islanders' rights and freedoms' (Cordes-Holland 2008: 2). Returning to Article 8, we can clearly make the argument that the failure to develop responsible and responsive mitigation policies may have the effect (especially in the long term) of depriving Islanders of their cultural integrity as well as resulting in dispossession of their lands. However, the complex emergence of anthropogenic climate change means that legal understandings of causation will prove very difficult in any litigation proceeding claiming negligence or harm (Cordes-Holland 2008: 10). Further, the future-oriented nature of some anticipated loss renders current legal action impotent. We cannot be compensated for what has not yet occurred.

While Article 8 places a responsibility on the relevant states to act, to prevent, and to redress, Articles 10 and 28 provide direction with regard to issues of relocation and compensation:

> Article 10
> Indigenous peoples shall not be forcibly removed from their lands or territories. No relocation shall take place without the free, prior and informed consent of the indigenous peoples concerned and after agreement on just and fair compensation and, where possible, with the option of return.
>
> Article 28
> Land, territories and resources equal in quality, size and legal status or of monetary compensation or other appropriate redress.

In the first instance then, forced removal is posited as untenable. The 'free, prior and informed consent' of community members is required for relocation to occur. The distinction between movement that is 'forced' and that which is 'consented' to is contentious. How much choice does one have when the land is rendered uninhabitable? At some point, decades down the track, 'consent' may present itself in the negotiations, but this consent is contextually bound, partial and cannot erase the conditions of violence and negligence preceding the 'decision' to relocate. Thus, Articles 10 and 28 include provisions for 'just and fair compensation', which will likely prove very significant in the coming decades.

HAUNTOLOGY: POLITICAL AUTONOMY AND THE FUTURE

In the case of the TSI, land as territory can take on a ghostly presence in the Australian imaginary in the sense that the region as a whole is frequently rendered invisible to Australians. This feeling of marginal presence is the product of colonial discourse which, as historian Jinki Trevillian has written, has long been expressed by Europeans who 'have also looked on Aboriginal people as ghosts, relics of the past, already dead and dying' (2008). In 2011, TSI Mayor Fred Gela appealed to the then Australian Prime Minister, Julia Gillard, to hear the calls for assistance coming from the Torres Strait: 'Help has been requested time and time again to assist in preventing the devastating effects of climate change, but unfortunately our own government has not come to our aid' (cited in Collerton 2011). Located at the tip of northern Australia, the region is rarely the subject of media attention or public discourse (as I will discuss in the next chapter). The feeling of being cast as invisible was immediately evident when I spoke with people on Thursday Island who referred to themselves as the 'forgotten people', repeating the ghostly colonial trope.

The loss of a specific Indigenous country, such as Saibai Island, will profoundly impact the identity of the local community, even as they retain broader rights to Australian citizenship. Before the seas rise to engulf these low-lying lands, the destruction of burial grounds for those still residing on the island draws our attention to the urgent need to advocate for adaptation funding that protects cultural rights. This is a cry for justice beyond that encapsulated in models of 'synchronic accountability' (Derrida 2006: 26). Government financing must contribute to the protection of sites of cultural importance, such as graveyards, which are the 'last dwelling place' (Dufourmantelle and Derrida 2000: 97). These sites are, for Derrida, more than just the final resting place for individuals, or sites of closure. Rather, he writes of the 'hospitality of the land and of burial' invoking an invitation at the event of finitude (Dufourmantelle and Derrida 2000: 85). In this sense, the place or site of burial holds the force of a welcome to the next generation; it invites engagement and respect for the past in preparing and moving into the future.

By engaging with the story of Oedipus and his process of deciding where he shall have his 'last dwelling place' (Dufourmantelle and Derrida 2000: 97), Derrida reveals to us the import of the burial site for 'defining home':

> The last resting place of family here situates the *ethos*, the key habitation for defining home, the city or country where relatives, father, mother, grandparents are at rest in a rest that is the

place of immobility from which to measure all the journeys ...
(Dufourmantelle and Derrida 2000: 87)

And so via this ancient narrative we are told of the ties that can form through land and death and the implications of this for carrying a sense of 'home' and culture into the future. This is not to be confused with a commentary on Indigenous burial practices or the significance of death in aboriginal cultures, which I am unqualified to offer. Instead, I am opening up a dialogue on the universal importance of one's burial on grounds we consider to be our 'home' and the potential for climate change to affect how this unfolds for some communities. Each culture has its own way of expressing this relationship. However, care with regard to cultural and spiritual matters of importance, such as burial grounds, may not be economically justified according to the rhetoric of 'priority investment' that the Australian government has indicated its support for in relation to the closure of remote communities in Western Australia.[7] The effects of climate change disturb both those resting in peace as well as those visiting the headstones. The failure on the part of the Australian government (and the state of Queensland) to adequately provide the infrastructure to protect these grounds suggests an indifferent attitude towards the future and past of this culture and its peoples.

Thus a haunted house could benefit from an ethical orientation that engages with and transforms the figure of the ghost. A haunted house is never present with itself: it is 'out of joint' (Derrida 2006). Derrida's 'hauntology' does the work of conceptual reorientation, taking into account the manner in which there is a 'disjointure in the very presence of the present, this sort of non-contemporaneity of present with itself' (2006: 29). The implications of this disruption to linear accounts of historical time is an irreversible, un-deferrable obligation to 'bear witness' to all that we inherit, including ghostly traces (2006: 68). The past is in the present, the future is in the present, the present is never in-sync with itself. Hauntology demands that we recognise a form of responsibility that fundamentally calls into question existing ways of being and doing predicated on neat distinctions between past, present and future. In this way, for International Relations theorist Jessica Auchter, hauntology can be understood to be an 'ethical responsibility' (2014: 25; see also Davis 2005). In particular, Auchter contends that the 'ghostly cannot be governed' and are therefore a force of resistance (2014: 29), something that is a major problem for the nation-state. The ghostly are ungovernable because they do not 'belong to the order of knowledge' (Davis 2005: 376).

In spite of this orientation outside of knowledge, or more accurately because of it, within Derridean thinking, the ghost is always already a disruptive force that cannot be assimilated. In the closing pages of *Specters of Marx*, Derrida asks if one could address *'oneself in general* if already some ghost did not come back?' (2006: 221). That is, without the traces of the past and the future, is there anything? Is there even the possibility of speaking? With the visitation of the ghost comes an imperative to enact justice:

> Not for calculable equality, therefore, not for the symmetrising and synchronic accountability or imputability of subjects or objects, not for a *rendering justice* that would be limited to sanctioning, to restituting, and to *doing right*, but for justice as incalculability of the gift and singularity of the an-economic ex-position to others. (Derrida 2006: 26)

If, as Mansfield argues, a 'future politics is a politics of ghosts' (2008), we must go beyond the calculative logic of cost–benefit analysis in our articulations of justice. This reorientation is disorienting (2008: 27). It puts 'our' dwelling in question. It forces us to recall, a task that Jenny Edkins argues is needed in order to unsettle 'sovereign politics' (2006: 114).

Practically speaking, what do the ghosts of the future demand of us today? The impact of climate change for Torres Strait peoples raises the spectre of regional autonomy. This is because what is absent once again is access to political representation, as well as the capacity to direct political decisions regarding the future of the region. Political action is fundamentally about the future which, as Derrida reminds us time and again, is always already 'present at the heart of experience' (1978: 118–19). The concept of autonomy – as a mode of self-government or internal control – is embedded as a right in the *United Nations Declaration on the Rights of Indigenous Peoples* 2007. Article 4 states that 'Indigenous people, in exercising their right to self-determination, have the right to autonomy or self-government in matters relating to their internal and local affairs.' Even if morally posited as a right, autonomy must be recognised or legitimated in order to be operationalised politically (Arthur 2001: v).

Politically, regional autonomy of some design may be of importance for the issue of climate change and its effects in the TSI, as well as the funding decisions that come to dominate responses to climate impacts. It may also prove significant if and when relocation becomes a reality for affected communities. However, it could be argued that planning which abides by adaptation guidelines such as Sustainable Livelihoods Frameworks could satisfactorily engage and empower the community in decisions concerning their future, putting in doubt the practical significance of autonomy.[8]

Discussions of the usefulness of autonomy remain unresolved in the region. Alongside the debates regarding self-governance, legal recognition of land has occurred. Mer Island (also known as Murray Island) is the site of the most significant challenge to Australian sovereignty, known as the *Mabo* decision (Mabo v Queensland [No. 2] 1992). Since this time all inhabited islands, except Hammond, have gained native title rights. In August 2013, the High Court of Australia ruled in favour of the Torres Strait Sea Claim, a legal decision which finally acknowledged the customary rights of Islanders over '44,000 square kilometres of sea country' (Butterly 2013). Western societies have historically privileged land over sea. Within this system of dualities, land is regarded as 'the theatre in which human interactions with the environment are played out' (Mulrennan and Scott 2000: 683). In contradistinction, Mulrennan and Scott tell us that 'Indigenous cultures commonly conjoin land and sea as continuous, integrated-"scapes", in conceptions of tenure, in resource management, and as bearers of social identity' (2000: 683). In an era of climate change, this integrated approach is sorely needed. It would add immense value to our political system and discourse to have such an ecologically sensitive and comprehensive viewpoint represented at the highest levels of government.[9]

Arendt's concept of politics as marked by plurality and contingency means that the inclusion of a TSI voice in the chambers of political debate and decision-making may promote the region and its interests into the future. This is particularly important in a climate change era which will, it is true, force all peoples to make changes to their lifestyles, but will impact some people more dramatically. In the case of the TSI, impacts exceed the concerns of lifestyle and go to the core of land, identity and belonging. Consultation, participation and political power are needed to ensure that in the event of communal relocation, culturally sensitive plans come to the fore. Before this, our national awareness of Islanders' adaptation needs is imperative. Formal political representation would also have implications for working towards the recognition of international Indigenous rights. Political inclusion in the form of representation is not an act of assimilation. Rather, calling for political recognition of this sort reminds us of the porosity of 'Australian' sovereignty and the manner in which it has never been, and never will be, present with itself. This sovereignty is haunted from without and within.

CONCLUSION: BORROWED GROUNDS

Current adaptation programmes in the Torres Strait Islands aim for community consultation and empowerment in the implementation of measures designed to keep people 'safe and in place'. These efforts continue in the face of a prevailing discourse in Australia which 'forgets' Indigenous peoples, their contemporary struggles, as well as the colonial history from which we emerged. The remote location of the TSI compounds this reality at present and all but ensures that the issues the community faces remain virtually invisible to mainland Australians. It seems both pertinent and urgent, then, to call for political autonomy in order to guarantee representation and a voice at the highest levels of the Australian government.

Relocation threatens an ethos nourished and sustained by a culture over generations. Sites of burial are the place of 'last dwelling' for the community (even some of those living away from Saibai). While the post-WWII relocation attests to the capacity for both cultural survival and transformation in different places, the trauma of the circumstances of this potential dispossession will result in a profound experience of loss. As the seas rise, what will disappear is not what colonial-era Chief Protector John Douglas described as a 'mythical' image of the Island (Douglas 1900). We will not only witness the loss of a geophysical territory subject to the dramatic changes of the climate. It would be moral negligence to fail to appreciate the implications that this open wound will have for future generations of Saibai Islanders, and Torres Strait Islanders, as well as for all Australians. As Derrida puts it, 'inheritance is never a *given*, it is always a task' (2006: 67).

The potential loss of material and intangible things leaves traces. These traces are the ghosts who haunt not only the descendants of those affected but the entire nation-state, and more than this, the globe. The possibility of absolute destruction is the ghostly threat unsettling the certainty of the promise of adaptation. Dominant Australian state sovereignty hosts the traumas of the past and those that will arise in the future. This is inherent to the constitution of the sovereign state. Ghosts from the future join others that remain from a colonial past. As such, it is a sovereignty that is challenged to confront its violent past and the ongoing traces its 'murderous, bruising origins' leave behind and project into the future, which is always already the present (Derrida 2006: 25). Only then we can dwell ethically and in the name of the future.

Chapter 6

HITTING THE GLOBAL NORTH: FROM CRISIS TO RECOVERY TO REBUILDING?

> 'The spectacle is not a collection of images; rather it is a social relationship between people that is mediated by images.'
>
> Guy Debord[1]

HARD RAIN

When Guy Debord wrote *The Society of the Spectacle* in the 1960s he was interested in the way that the economic system of a political community managed to hold considerable power. He argued that this was fostered through the construction of an 'industry of the image'. This industry sought to reinforce the particular values and identities of the ruling class, whether communist or capitalist, totalitarian or democratic. Within capitalist societies the model of such a spectacle was 'diffuse' and emphasised an 'abundance of commodities' (Crary 1989: 105). That is, the visual realm, what Debord called the 'spectacle', propped up the interests of a capitalist economy through the proliferation of images of consumption for the population to feed on. In totalitarian regimes, the spectacle was understood to be more coercive and emphasised a state of 'permanent violence'. By the late 1980s, he had collapsed these two models, referring to an 'integrated society of the spectacle' where commodification and violence worked together (Crary 1989: 105).

In the Global North, climate change is something to be consumed through the ever-diversifying media platforms available to us. It is a spectacle that invites its audience into a relationship with others. The social relationship that is enabled can take many forms: from those that are motivated by compassion and altruism, to those that dehumanise and exclude. Regardless of what cultural values arise in the formation of this social relationship, the politicised consumption of disaster has ethical consequences for how we in the Global North understand and engage with the crisis that climate change poses, particularly if it involves chronic displacement from one's home. In the Global North, we are suffering the plight of dwelling within a Heideggerian register. Not only do we fail to take care of the world we are thrown into, but we remain committed to technologisation, or earth-as-resource, as an answer to our social and environmental ills. Moreover, extrapolating from Levinas we can see that we routinely delude ourselves with the belief that our possession of the world is secure.

As I have demonstrated throughout this book, climate change follows well-worn paths that tend to reinforce patterns of social, political and economic disadvantage. Indeed, climate change will give rise to 'a politics that already divides the population between the prioritised and the expendable' (Mansfield 2008). This chapter focuses on the politics of expendability and priority as it has played out during the political and media spectacles of two major disasters in the wealthy West: the 2011 Queensland floods in Australia, and the 2005 event of Hurricane Katrina in New Orleans. A politics of expendability and priority is one that is concerned with constituting the value of human life in accordance with a specific articulation of who is *entitled* to dwell; whose dwellings are considered important in the recovery and rebuilding efforts.

Analysing the discourse of 'dwelling' as a political construct circulating in the Global North's consumption of climate crises reveals the manner in which the question of who has the right to be at-home, that is, who is permitted to dwell, is entrenched in histories of social exclusion. We saw this in the previous chapter, where the dwelling of Indigenous Australians must continually negotiate the challenge of neo-colonial power relations. Specifically, there is a *politics of domicide* unfolding in such settings where the specific needs of marginalised peoples, including refugees, racialised minorities, queer communities, and other vulnerable groups, are either neglected or excluded from disaster response and recovery policies (Gorman-Murray et al. 2014; Cresswell 2008; MDA 2011).

'Domicide' is a term coined by Canadian geographers Porteous and Smith. It refers to the 'deliberate destruction of the home against the will of the home dweller' (2001: 3). Even as the original authors posited that 'natural disasters' could not qualify as 'deliberate', updated articulations of domicide have challenged this by emphasising the social and political nature of all so-called 'natural disasters' (Gorman-Murray et al. 2014). I argue that everyday domicide is able to go under the radar in places such as the United States and Australia because climate disasters are increasingly inscribed in a vocabulary of the 'new normal'. As such, following the immediate impact and response to disaster, it is assumed that all citizens will take personal responsibility for their progression from crisis to recovery to rebuilding. Resilience in this context is a thoroughly individualised concept used to explain why some are able to adapt while others fall by the wayside. In this chapter, I argue that narratives of trauma and loss circulate in ways that encourage us, as 'media witnesses' (Frosh and Pinchevski 2009), to either connect with the Other who is suffering, or to dehumanise the Other (and therefore blame their 'failure' to adapt on personal inadequacies). This is particularly so when the impacts of disaster result in chronic homelessness or displacement within our cities and towns.

THE 'NEW NORMAL'

Climate crisis has taken on the label of the 'new normal'. Indeed, the World Meteorological Organization has called for a shift in the 'climate baseline' to reflect the 'new normal'. What is 'new' is the increased frequency of the idea of extreme weather events globally (Hannam 2014). The UK newspaper *The Independent* included the following headline as early as August 2013: 'Extreme heatwaves are predicted as the new normal for British summers by 2040' (Bawden 2013). This discourse of the 'new normal' is recognisable in both politics and the media in the Global North. It works to expand the field of vulnerability to climate disasters in such a way that exposure is assumed to be a socially equalising condition. With the emergence of climate crisis as the norm, citizens are positioned both as spectators and potential victims. This is achieved through the proliferation of images and narratives of suffering. Stories of trauma and loss associated with environmental disaster are repeated, and remind the 'media witness' that they could be the next in line to lose everything.

Spectacle

To open this chapter, I drew upon Debord's notion of the spectacle. My interest is in the way Debord ties the spectacle to the generation of different forms of social relationship. What is the significance of Debord's reflections on the media today in relation to climate change? This question is important in a context in which it is reductive and impossible to draw neat causal lines between the mass media/new media/social media and the political 'ruling class' in the manner that Debord did. According to writer Jonathan Crary, Debord was fundamentally concerned with the ways in which the society of the spectacle led to the 'management of bodies ... [and the] management of attention' (1989: 105). Based on this interpretation of Debord's work, we have two major points to be explored. Firstly, the media spectacle authorises and privileges, or alternatively discards and marginalises, bodies. In the event of climate disaster, it is through a focused and sustained narrative of trauma and loss that specific privileged bodies are recognised as vulnerable in the first instance and 'grievable life', in the second instance (Butler 2004). Other bodies are also caught up in the media spectacle, but are marked as 'disposable', 'ruined', 'irreparable' and so on (Giroux 2006, 2008).

The infamous contrast during the New Orleans Hurricane Katrina disaster, of the coverage of a white couple 'finding' food supplies and a black man 'looting' the local grocery store, immediately comes to mind (Jones 2005; Cresswell 2008). The desperation of the white couple stumbling upon supplies necessary for survival in tough circumstances sits alongside the criminalisation of the very same behaviour for the black man. When such bodies are 'caught in the act' they are immediately cast in well-worn racist categories. Adding to this already distressing scenario is the way in which it is precisely these acts of desperation, and the narratives that stitch them into broader discourses, that is readily and perhaps excessively consumed by spectators across social, new and traditional media forms. Here, the image turned into a critical intervention in the racial politics of disaster response and recovery, but more frequently images circulate without such critical purchase.

In another context, media theorist Anita Biressi has written about the ways in which particular forms and expressions of bodily trauma come to be displayed and consumed through contemporary media (2004: 335). Embodied trauma and the frenzied consumption of 'personal pain and suffering' saturate today's media platforms (2004: 335). This is a pertinent observation when it comes to the coverage of climate and 'natural' disaster. It is through images of loss, that is, loss of home, livelihood, control, lives and security, that social relationality is cultivated for some while denied to others. In other words, the culture of the spectacle produces the climate crisis as the 'new normal' available to consume on demand. As a part of this, it contributes to the generation of social hierarchies in which pre-existing forms of political division are compounded.

As I will demonstrate in this chapter, support for some communities fosters forms of rehabilitation, resilience and recovery or, alternatively, reinforces modes of social, political and economic marginalisation and moral dispossession. Victims and audiences alike are drawn into these discursive positions. The narrative options available map onto economic and social differences, particularly those of race and class, but not exclusively. In sum, the visual economy of climate crisis privileges some bodies while marginalising others. This authorises the trauma of some while denying the recognition of others, or perhaps put more cynically, allows the representation of the latter until the audience has gorged on it and must then purge themselves of its impacts.

The second major point to be taken from Debord refers to the concept of time or attention. The spectacle sacrifices the past and future for an unrelenting emphasis on the present. An implication of this includes the failure to engage historical context in the events unfolding. The coverage of climate disaster both supports and undermines this. Because climate-related events are frequently narrated as 'natural disasters', the historical aspects of anthropogenic climate change are purposefully denied. It has been reported that in the days following the 2012 Superstorm Sandy in New York, at least ninety-four newspaper stories failed to mention either 'climate change' or 'global warming' (Leber 2012). In other words, 'natural disaster' is seen to come out of nowhere, a freak occurrence that hits without warning and without reason. Due to this emphasis on chance and unpredictability, structural factors which contribute to impact and recovery and rebuilding efforts, such as class, are also often not adequately addressed in either policy formation or the media (Gorman-Murray et al. 2014). However, the future is often part of the discourse of climate change with images of apocalypse increasingly prominent. The spectacle of climate crisis collapses current catastrophe with future-oriented discourses of disaster and even species annihilation. Yet in a move that at once accommodates or minimises and exemplifies the threat contained in such discourses, crisis comes to constitute the 'new normal'.

Plight

It may be tempting to suggest that recognising the impacts of climate change as the 'new normal' is a progressive and important development. After all, in parts of the Global North such as Australia and the United States, human-induced climate change is still considered hoax science. It is cast as the stuff of science fiction, not social reality. However, something else is happening when climate impacts are normalised. The 'new normal' needs to be understood as such because the threats that are and will arise must also be responded to. In the world of the 'new normal', catastrophe will follow catastrophe, and the future will be experienced by citizens as perpetual and unrelenting exposure to crisis and uncertainty. Drought, storm surges, flooding, rising seas and cyclones will mark the earth's territories indiscriminately. Worse still, as the historically wealthy states in the global context, the Global North is cast as the potential site of flooding by 'climate change refugees', pouring over the borders of the Global South. The Global North is cast as doubly vulnerable; susceptible to danger from within and from without.

In the wealthy West, this process of normalising climate change is accompanied by both a security discourse and an individual responsibility narrative. In turn, this produces a subject- (or nation-state-)in-crisis who must swiftly act to rebuild the status quo. This security discourse promises to minimise threats. It is easy to have sympathy with this reaction. Surely we all want security. Underpinning this reaction is a desire to control and exclude. The baggage of security discourses means that when it is invoked with reference to climate change we close down questions of justice and the burden of the future as a responsibility towards Others. While the crisis of climate change should provoke an existential crisis, the severity of the situation is obscured. This is not a crisis of existence, but one of control. The remedy is in reinstituting borders, walls and laws, not reflecting on our relationships with the environment and others!

In this sense, the spectacle of climate crisis that the wealthy West consumes reinforces dualisms such as culture and nature, us and them, self and Other. The order promised by a security response (securitisation) conforms to the delusions of a calculative politics that claims to be able to contain the uncertainty of the future in advance. What we need instead is to engage in a process of critical media and political consumption. We must counter the universalisation–securitisation–normalisation discourse with a commitment to understanding that 'crisis' requires from us the act of decision, not calculation. While we are all mortal and vulnerable as human beings, we are not all exposed to danger in the same ways. The climate crisis must remind us, those of us watching, witnessing the events from afar, that we dwell-with Others who invite us to respond to them with care and compassion.

The capacity to remain open to the Other's suffering is challenged in a social context in which the dominant discourses of our individualistic culture encourage us to act with self-interest only. Derrida, glossed by Sokoloff, offers us the concept of the 'interrupted subject' as a way of trying to explore the ethical possibilities that are generated when we refuse to take up the assigned role of closed-off neoliberal subject: 'An interrupted subject is not the negation of the subject but one newly conceived. That is to say, the subject who decides must be oriented to doing justice to the Other as opposed to maintaining its self-identity' (2005: 345). What is interrupted is the idea that we are separate. The intersubjective is a space of vulnerability. But this condition is what allows for connection and care for others. Without an embodied appreciation of vulnerability we cannot engage with the experience of another who is also vulnerable by nature of being a sentient being. There is a tendency to assume that constitutive vulnerability or exposure equates with a state of perpetual weakness. When this thinking asserts itself as truth, weakness is then the base position which gives rise to experiences of crisis. The solution or remedy is wrapped up with the demand for strength and security. The characteristic response to this perception of weakness is a reactive gesture of closure, coupled with techniques of control which seek to reduce or avoid the perceived threat. A response to vulnerability which immediately folds back into weakness can only enact a politics of security. That is, the 'subject-in-crisis' sees the Other as a potential threat to be expelled.

The 'interrupted subject', as opposed to the 'subject-in-crisis', opens onto the Other without the reflex of closure. The 'interrupted subject' is one that is exposed. But this exposure orients it towards an understanding of the importance of mortality; my own (as per Heidegger) and that of all Others (as per Levinas and his notion of the Third). Recognising the Other as a mortal reminds me of my responsibility towards that person. Crisis as norm loses this crucial aspect of interruption as a reminder of fragility and a call to accept one's concern for the welfare of another. Both deploy the logic of crisis, with the former emphasising threat and disintegration and the latter interruption and responsibility. Crisis should not be regarded as 'normal'. Instead, it should retain its extra-ordinary status.

In Derrida's account of decision, it is precisely the discrete human agent that is undone. The crisis unravels the self, a point recognised in the original Greek understanding of *krisis* as decision. A deconstructive reading does not then do the work of rebuilding the identity of the person. Instead, the 'interrupted subject' is the one able to decide and act with the knowledge that the outcome of decision is not the restoration of an undisturbed identity. Something different or new will emerge. Decision is only possible because I am part of a world that exposes me to risk and forces me to live in a context in which the effects of my decision are to some extent incalculable. The point of this reorientation of decision is to appreciate that what spurs action is not an inherent capacity in the human agent but the force of the demands the Other places on me. Ethics, too, is not something I opt in and out of. Ethics comes from the Other.

My critical approach to the emergence of the new norm of climate crisis is not the same thing as failing to acknowledge the radical risks we face. The crisis of climate change – as irruption, turning point, and severe – brings us to the 'brink' of life (Irwin 2008: 17). My point is to help to redirect our response (and before this, it is to understand the conceptual foundations we are working with). Crisis, in the emerging register, calls for mere management of costs and benefits, or risks and possibilities, and the securitisation of our dwelling place. Moreover, it demands that anyone affected by disaster promptly take personal responsibility for their recovery and rebuilding. But this sense of rebuilding is restricted to the dictates of our neoliberal culture: we are asked to rebuild in the name of economic and political security.

For philosopher Ruth Irwin the fact that climate change exposes us to the possibility of extreme loss means that we are also situated in a way which offers the possibility of making a decision that recognises something other than the status quo. Drawing on Heidegger, she writes:

> Heidegger's approach to the zone that threatens to draw modern culture to an untimely close, is not entirely pessimistic. He regards the 'line' or boundary of the planetary conditions for human life as generating a new 'beginning'. That beginning (or the contempla-

tion of the ending) generates the conditions for rethinking what it is to be meaningfully human. (Irwin 2008: 133)

Whether it is individual mortality or collective (species) annihilation, a consideration of temporal contingency and impermanence reinvigorates questions concerning the meaning of life. Crisis can give us pause to consider new ways of co-existing with one another and the earth as a whole. Crisis is not to be reduced to the embodied alertness to danger and disaster, or the perpetual consumption (via media witnessing) of trauma. Crisis demands that we consider how to create change that is meaningful; it demands that we act. Climate crisis as spectacle may generate action but only for a short period and only in the interests of the privileged.

For Debord, the society of the spectacle led to a deep slumber. Indeed, society was 'in chains' and 'expressing nothing more than its wish for sleep'. The spectacle, he concluded, 'was the guardian of that sleep' (1967: 7). In the age of climate catastrophe, society is haunted by recurrent nightmares. In the grip of terror, we turn inward. Debord's notion of the spectacle as contributing to a society incapable of staying awake, and hence incapacitated in the realm of social and political action, has a place in understanding why inertia seems to have set in politically with reference to climate change. It is analogous to Heidegger's diagnosis of what comes to constitute the 'plight of dwelling'. Indeed, this is our spectacular plight.

Wound

The climate spectacle is not a call to action in any ethically transformative sense. Today, this is complicated further by the contribution of what Mark Seltzer refers to as a 'wound culture' which results in 'the convening of the public around scenes of violence'. In Seltzer's words: 'The convening of the public around scenes of violence ... has come to make up *wound culture*; the public fascination with torn and open bodies and torn and open persons, a collective gathering around shock, trauma and the wound' (cited in Biressi 2004: 345). Wound culture goes hand in hand with both the production of the 'new normal' and the discourse of universal vulnerability which comes to accompany climate change. Trauma as spectacle is not transformative of our social relations. The task is to turn this attention on trauma towards a political intervention aimed at social and ecological justice; and to return the significance of interruption, irruption, disequilibrium, decision and risk to the crisis and act accordingly and in the interests of the future. Crisis is not the unfolding of the ordinary. Its extraordinary character exemplifies the significance of a political decision and enlivened ethical action which takes account of what is at stake, rather than reducing crisis to a routine threat and assimilating it into the logic of security.

UNDOING UNIVERSAL VULNERABILITY: DOMICIDE AND SOCIAL MARGINALISATION

There is little doubt that global climate change renders each of us potential victims. In the face of severe weather we are exposed to the possibility of losing everything we have worked for and value. Climate change reminds us of the constitutive vulnerability we embody as mortal beings on a planet that is indifferent to our individual and collective passing. However, vulnerability is a social phenomenon, not merely an inherent characteristic. That is, while we may all feel a sense of victimhood, some people occupy positions of greater vulnerability. Indeed, climate change is not something which happens outside of existing social and political structures. Consequently, it is 'embroiled within issues of inequality and human rights' (Maldonado et al. 2013: 602).

Vulnerability is also influenced by the sorts of environments we live in. Cultural theorists Laurie Anne Whitt and Jennifer Daryl Slack argue that 'Geographical and ecological features of community are rarely incidental to political and cultural struggle: they contextualise – enable and constrain – relations of power' (1994: 6). In each of the examples I explore below, the site of disaster is significant to understanding disaster preparedness, response, and recovery, as well as its inequitable character. It is not just bodies which come to be caught up in the spectacle of climate change, but also the environments they are situated in. Some environments are privileged and rebuilt over others, and even within a particular setting, some regions elevated while others are left in states of ruin or disrepair following disaster: a politics of domicide. Social and ecological decay occur simultaneously, reiterating the link between the two. Indeed, 'often a sense of home or meaningful dwelling involves the residential house and neighbourhood simultaneously', offering a type of 'ontological security' (Gorman-Murray et al. 2014: 244). Domicide, then, involves the loss of material structures but also those of affective, social and cultural importance. Accounting for these intangible losses, domicide encompasses the 'erasure of the spatiality of home not necessarily the destruction of property', as such (Ó Tuathail and Dahlman cited in Nowicki 2014: 789).

BRISBANE, QUEENSLAND, 2011: 'MADE OF TOUGH STUFF'

In Chapter 5, I explored the urgent and potentially devastating impact of climate change in the Torres Strait Island region. King tides affect the region on an annual basis, leading to a possible increase in health concerns such as water-borne diseases, as well as interference with local aquaculture which sustains the communities. Vulnerable islands include Masig (Yorke), Poruma (Coconut), Warraber, Yam, Saibai and Boigu, which are less than 1 m above sea level. Calls for Federal government commitment to the construction of sea walls were heard again without success in March 2012, with the Torres Strait Council arguing that work on six islands would cost approximate $22 million and provide an extra twenty or thirty years for the islands before relocation need to be considered (T. Cohen 2012).

The chronic nature of climate change impacts in the Torres Strait contrasts, at least in the realms of political response and media representation, with the time-bound nature of the floods that affected the state of Queensland in 2011 (barring the Torres Strait Islands). In late 2010, flooding began to affect large parts of Queensland. By mid-January 2011, 75 per cent of the state was declared a disaster zone (Hurst 2011). The capital city of Brisbane was dramatically hit for the first time in several decades, with the banks of the Brisbane River breaking. This caused significant damage to housing, transport and the general infrastructure in the city. The damage to the environment was coupled with the tragic death of thirty-five people across the state. Many Australians lost everything they had worked for: their businesses and their home. Insurance companies refused, in some cases, to cover flood damage. Political responses to this event were immediate and comprehensive. The Prime Minister at the time, Julia Gillard, addressed the public alongside the then Queensland Premier, Anna Bligh, both expressing condolences and outlining measures on behalf of government aimed at assisting with the clean-up and rebuild.

Extensive and continuous media coverage portrayed many facets of the community atmosphere, taking on the role of framing embodied human vulnerability, of making life 'visible in its precariousness' (Butler 2010: 51). The emotional, social, economic and cultural impacts of the Queensland floods at both the personal and the national level are significant and I do not wish to brush them aside or be dismissive as I offer this critical engagement. However, I am interested in the distinct ways in which the examples of the flooding in Brisbane, and climate change in the Torres Strait, play out politically. What does this tell us about the visibility of vulnerability in a neo-colonial Australian setting? Moreover, even within the affected regions of Queensland certain groups of displaced Australians were both more vulnerable prior to the disaster, and differentially treated or insensitively engaged with in the response and recovery efforts (Sevoyan and Hugo 2015). For example, the failure to care for the needs of transgendered peoples with regard to emergency shelter and housing has been reported as a significantly marginalising effect (Gorman-Murray et al. 2014: 246).

It is useful to juxtapose the cases of the TSI and Brisbane because both sites rely upon the co-operative action of the Queensland state government and the Federal government in the wake of disaster or climate hazards. In this sense, at the political level, the two cases illuminate the manner in which national responses are affected by the implicit or explicit 'value' inhering in the location and peoples. That is, these two examples tell us a story about whose dwelling we care for and whose we are willing to sacrifice or ignore. As media witnesses we are brought into an intimate relationship with the suffering of some, but not others. The locations of the TSI and Brisbane are valued differently, according to the economic potential that is perceived to inhere in each location. While it is a truth we would prefer to deny, this impacts upon the way that political responses unfold. The Queensland floods threatened an urban, consumerist city, on the mainland of Australia. The Torres Strait is funded by the Queensland government and has very little independent income-generating potential. While the Federal government has an active interest in the latter region for national security and quarantine purposes, its primary base on Thursday Island is not currently at risk. In other words, when disaster strikes in the TSI the costs are high and the benefits perceived to be low. Added to this is the fact that the region has no major industries that the state or Federal government are invested in, and thus there is minimal potential through future prosperity to reverse the (perceived) economic burden of material rebuilding. On the other hand, the reported cost of the Queensland floods exceeded $5 billion, with annual profits from coal production lost on top of this (The Age 2011).

Former Prime Minister Julia Gillard referred to the floods as a 'national tragedy', which resulted in 'the awful loss of human life' but had not broken the spirit of Australians, with flood-devastated regions 'moving from crisis to recovery to rebuilding' (Levy 2011). This progress was assisted by federal and state government funding. A national flood levy was introduced to raise the $5.6 billion bill (Levy 2011). The Queensland government supplied $233 million to be distributed to individuals and families left homeless. Then Premier Bligh remarked that this money would help to 'provide some certainty for people whose homes have been utterly destroyed, who are literally rebuilding their lives after these events' (Martin 2011). Here, visible precariousness is responded to appropriately by government with social justice measures enacted to assist those left with little. However, this was not universal. The Multicultural Development Association (MDA), an organisation that provides support and information to recently settled refugees and people from culturally and linguistically diverse backgrounds, noted that emergency warnings were not provided in appropriate languages for its clientele to access (MDA 2011: 2): 'While the floods have been a traumatic time for many Brisbane residents, the trauma for MDA clients was exacerbated in many cases by reminders of previous traumas and homelessness as well as helplessness from lack of information' (2011: 8). Moreover, the organisation identified the long-term effects of the floods, noting that there was a grave risk of 'homelessness of many refugee families' (2011: 6).

Queensland's 'Local Refugee'

The Queensland floods demonstrated the possibility of co-operative government efforts aimed at reducing the suffering experienced when one is left homeless. Moreover, the spectacle of the floods elicited a social relationship between the victims and the audience viewing from our lounge rooms. An empathetic connection was enabled through the use of the term 'refugee' to bring home the involuntary and indiscriminate nature of the conditions experienced by fellow Australians. Homeless Queenslanders were termed 'refugees' across Australian media outlets. Headlines included 'Tent metropolis for flood refugees'

(Tranwith 2011), 'Evacuees feel like refugees in their own city' (ABC 2011), with one article 'PM plays with kids at evacuation centre' telling us that 'Ms Gillard sat with two flood refugees' (ninemsn 2011).

The extension of the term 'refugee' to refer to people left homeless following flooding highlights the manner in which vulnerability to natural calamities is regarded as universal. The effect of this is to even the playing field: the media witness looks upon the crisis unfolding and realises that this could happen to them too. Even as this may elicit compassion and action, these possibilities rely upon the potential of seeing ourselves in the place of the victim. Thus, the climate victim must be seen to be 'one of us'. The discourse of the local refugee – 'We're all Queenslanders now', one headline shouted (Sydney Morning Herald 2011) – contrasts dramatically with the visual representations of the global phenomenon of climate migrants. Climate migrants are viewed as invasive and homogenous. Moreover, this discourse of the local refugee is blatantly inappropriate given public perception and government policy regarding refugees who arrive in Australia fleeing political persecution, as well as the aforementioned inadequacies of the government's response to refugee communities in the region.

Further, while we were invited to reflect upon the unjust nature of climate/natural destruction when it hit our fellow citizens, soon after these events a news story claiming that 'fifty million "environmental refugees" will flood into the Global North by 2020' drew on the familiar language of waves to frighten us, in the Global North, of the possibility of migration from without (The Australian 2011). The spectacle of the local refugee reveals a sense of being wounded and exposed. Yet the trauma of the foreigner refugee remains silenced and excluded from Australia's political landscape and moral vocabulary.

The limits of our empathetic witnessing are evident within the country, not merely in relation to foreign asylum seekers or migrants. Shortly following the devastating Queensland floods, in July 2011, the Torres Strait Islands Mayor, Fred Gela, wrote a letter to Prime Minister Gillard with the following message: 'We cannot afford to keep waiting forever. Failure to act on desperately needed adaptation measures in the Torres Strait puts Australia at risk of being the first developed nation with internally displaced climate change refugees' (Collerton 2011). In this call for help, the term 'refugee' operates differently again to the discourse of the 'local refugee' evident in the coverage of the floods. When Mayor Gela utilises the language of the refugee he is bringing our attention to a systemic political failure at all levels of government in relation to the region. To invoke the figure of the refugee signals that our government has knowingly failed to act. These 'refugees' will be internally displaced and thus not technically 'refugees'. Yet they would also be forcibly expelled from their homelands and indeed, one may argue that should this happen they would be victims of domicide. Repeated calls for a sea wall have fallen on deaf ears, while Torres Strait Islanders deal with king tides, inundation and the threat of rising sea levels. The funding extended to victims of the Queensland floods aimed, as Premier Bligh noted, to provide some basis for security and certainty. In the case of the Torres Strait no such engagement is secured. There is no promise to extend care.

What is revealed in these two examples is the narration of climate impact as differentially recognised: the greater the impact of climate change, the less textual space the issue takes up, the less it registers as a concern for Australia's construction of its 'home'; its national ecology. This reinforces Whitt and Slack's (1994) assertion that relations of power are invested in geographical and ecological contexts. On the one hand, the Torres Strait received a small spate of media stories in 2006 and early 2012 but very little overall media attention (the Australian Broadcasting Corporation covered the story I highlighted above). The region has had very little government funding and climate change only just registered as a policy issue. In fact, it did not feature in the *2009 Queensland Government Smart Report* into environmental concerns (Levin 2010). On the other hand, the event of the Queensland floods dominated air time and quickly prompted a flood levy (Levy 2011), as well as, rightly, government reports into insurance and disaster response and management, as well as state government funding (Martin 2011). Following the particularly devastating conditions in the town of Grantham some members of the community were relocated to higher ground with the assistance of the local council.

Part of the difference in response could be put down to the urgent nature of the flooding in Queensland and the slower, more long-term nature of the situation in the Torres Strait, but this explanation is not enough. Climate change raises questions that a democratic country like Australia must confront; concerns which go all the way from the level of government and the guarding of sovereign borders, into the more intimate spaces of the 'home', where individual behaviours have a collective impact upon the earth and its functioning.

KATRINA, NEW ORLEANS, 2005: 'WE ARE NOT REFUGEES'

The Queensland context reveals a number of different ways that the label of 'refugee' has been taken up. Firstly, it has been deployed as a political strategy by marginalised groups in the TSI. Secondly, the term has operated in media rhetoric as a way to illuminate 'universal' vulnerability and victimhood and to elicit empathy from audiences. Following the devastating impact of Hurricane Katrina in New Orleans, the label 'refugee' came to be associated with the victims left behind in the now infamous Louisiana Superdome. A cursory 'Google image' search for 'Hurricane Katrina refugees' reveals masses of pictures of stranded victims in sprawling, disorganised, camp-like situations. However, unlike in the Australian setting, in the United States the language of the 'refugee' drew immediate criticism and was the subject of a debate pivoting around themes of race, class and citizenship. Ultimately, the appropriateness of the term was rejected. It could not serve as either a political tool or a suitable descriptive label for the conditions of despair locals found themselves subjected to during and after the disaster.

However, once again the use of this term reveals to us the racially informed construction of the relationship between dwelling, place and belonging. American anthropologist Adeline Masquelier writes that the introduction of the label brought to the fore a longstanding and destructive 'racialised discourse' (2006: 736). This was evidenced in the differential coverage in the media of 'white "evacuees"' as opposed to 'black "refugees"' (2006: 739). This distinction and its embeddedness in pre-existing prejudices had the virulent effect of casting black New Orleanians as equivalent to non-citizens, an assignation that was quickly challenged by leading African American spokespeople. At issue was the conflation of black Americans with weakness and poverty. Masquelier notes that this contestation over meaning led to the adoption of an alternative vocabulary such as 'victim' and 'evacuee' to describe the conditions of disadvantage and need (2006: 736). The political message is clear: when black Americans are cast as refugees, they are effectively excluded from the body politic of society and as such from the rights and entitlements of citizenship. In turn, the local media witness is encouraged to view fellow Americans as outsiders. Black Americans begin to blend in with other popular representations of global 'refugeeness'. As one Katrina victim put it:

The image I have in my mind is people in a Third World country, the babies in Africa that have all the flies and are starving to death ... That's not me. I'm a law-abiding citizen who's working every day and paying taxes. (Masquelier 2006: 737)

What the rhetoric of the refugee does highlight is the status of an event as a crisis in the strongest sense of the term. The use of the term 'refugee' immediately tells us that what we are witnessing is devastating. We are reminded that a crisis is unfolding.

Putting aside the politics of the label, when someone is determined to be a refugee a set of moral duties are invoked. The duty to provide refuge or shelter to the victim is paramount. In the absence of thorough domestic legal channels for sheltering victims after the hurricane, this moral principle was left to the community of neighbours to extend in the aftermath of the disaster. The next two sections outline the racialised discourse of refuge which unfolded in the immediate post-Katrina landscape, followed by a brief overview of the larger failure to provide homes for those evacuees seeking to return to their neighbourhoods in the months and years that followed.

Crisis Management

Rather than elicit empathy or compassion, scenes of widespread disaster can produce a failure to extend care. This is done when the media witnesses distance 'themselves from the victims, either by deciding that they are not "innocent" or by situating them firmly within the confines of an outgroup' (Sommer et al. 2006: 50). This is a psychological 'strategy for coping' with the loss of a sense of the world as ordered and reliable (2006: 49–50). The promise of empathetic relations in the recognition that we are part of a shared world, or at the very least bound to the same conditions of contingent existence, is only partially enacted. Following Hurricane Katrina, victims unsuccessfully sought refuge across the Mississippi River in a town called Gretna. Hundreds of evacuees encountered local Gretna police as they attempted to cross the bridge connecting the town to greater New Orleans. The Mayor of the town stated that Hurricane Katrina 'was not a 9/11 tragedy with good-heartedness all around. You had anarchy and civil disobedience' (Inniss 2007: 332). Other cities were equally hostile, with Houston's elected representative proposing a law which would permit the removal of evacuees who were charged with a criminal offence while residing in the area (2007: 331).

The extension of hospitality in times of crisis is an opportunity to foster forms of social relationality previously absent or even socially discouraged. Yet when a 'wound culture' dominates, the possibility of transformative social relations is diminished. Consumption of trauma and loss is detached from a sense of its consequences or a call to engage ethically. Wound culture as a spectacle emphasises the immediate, and as such is unable to accommodate drawn-out or chronic conditions, such as displacement, dispossession and migration. This means that after the initial saturation, coverage of the implications of disasters lessens, and the long-term effects of disadvantage go unnoticed or worse, deliberately ignored. Chronic immobility, displacement and homelessness are cast as individual failings, not forms of structural violence (Giroux 2006; Beckett and Herbert 2012).

In other words, by emphasising the 'now', the context and contributing role of structural and institutional marginalisation is obscured in favour of a focus on individual responsibility. This has a twofold effect. In the first instance, the form of suffering able to be consumed without consequence is that of the privileged who, while enduring an undeniable hardship in the midst of a disaster, are better resourced and able to 'manage' and return to a state of normality within a 'reasonable' amount of time. Indeed, the capacity of the privileged to return to a relative status quo comes to set the benchmark for 'resilient' communities and individuals. In the second instance, the inability of the poor and/or racially marginalised to readily bounce back is cast as a problem not for the society, but of the individual and, equally problematic, the community they are part of. One Florida-based radio host used the dehumanising language of the 'parasite' to frighten listeners in relation to the possibility of evacuees seeking shelter in their communities (Giroux 2006: 176).

For Giroux, the systematic refusal to acknowledge the dignity of Katrina survivors of African American background is less indicative of a psychological coping strategy in the face of instability, and more a reflection of the extreme violence that is part and parcel of living in a neoliberal society. Giroux argues that:

> Something more systemic and deep-rooted was revealed in the wake of Katrina – namely, that the state no longer provided a

safety net for the poor, sick, elderly, and homeless. Instead, it had been transformed into a punishing institution intent on dismantling the welfare state and treating the homeless, unemployed, illiterate and disabled as dispensable populations to be managed, criminalized and made to disappear into prisons, ghettos, and the black hole of despair. (Giroux 2006: 175)

In relation to Hurricane Katrina, this management of underprivileged bodies started with the bombardment of stories and images of displacement and loss. Visuals of the camp-like situation at the Superdome saturated the media. This was followed by narratives of squalor and violence, of desperation and despair. As Biressi highlights, in the role of media consumer, it is still the human body that is the site of trauma; only this becomes public rather than private business. The traumatised body of the Other marks both the possibility of engagement and its limitations. Giroux is correct when he argues that in a neoliberal society the poor are 'excommunicated from the sphere of human concern ... rendered invisible, utterly disposable' (2006: 175). However, this is a complex process which first involves their inclusion in the politics of crisis. That is, in a wound culture, the poor are not immediately cast aside, but are instead on display as both evidence of the injustice of destruction, and a warning of the horrors of individual deprivation.

The individualised nature of this presentation of loss (which only ever takes on a collective dimension when presented as endemic and inherent to a cultural community) has two specific effects. Firstly, the structural factors which create and sustain vulnerability, poverty and disadvantage are erased or ignored. Seltzer's contention that wound culture is pathological is confirmed in that the public is encouraged to gather around these 'public scenes of violence', with little chance that the coverage will advocate for social and/or political transformation. Secondly, the accompanying experiences of chronic (as opposed to immediate and short-term) displacement, dispossession or migration exceed the limited attention that the 'spectacle of the image' of trauma can accept.

A Place of No Return

As American journalist Katy Reckdahl writes, the reception in adjoining cities and towns of Katrina evacuees in the days and weeks following the disaster raised a profoundly uncomfortable question, 'Is there an ideal evacuee?' (2006). This absolute refusal to welcome stranded peoples is simultaneously a refusal to reflect on the conditions of one's own dwelling: to answer for one's dwelling and to understand the relational nature of our being in the world. Troubling also were the strategies of exclusion employed in New Orleans towards residents in numerous disadvantaged areas of the city who sought return following the disaster. With regards to this problem, Lolita Buckness Inniss (2007) addresses the legal 'right to return'. She asks why 'poor, black Katrina victims' have not returned to the city. Her answer identifies an undeniably discriminatory approach to repatriation and resettlement, and gives four reasons as to why this is the case. The first is poverty. Many of those chronically displaced are simply unable to afford to migrate back, or had been in a situation in which they have had to migrate multiple times in search of shelter and employment. Secondly, barriers related to housing, particularly insurance, were insurmountable. Many poorer families and individuals did not have comprehensive (or any) insurance. In fact, those residing in the Lower Ninth Ward of New Orleans were informed by the Federal Emergency Management Agency, before the hurricane, that their location was protected by the levees and as such insurance was not necessary. Thirdly, with the rebuilding of New Orleans the renovation of housing led to an increase in rental pricing across the city. The Brookings Institution estimates that rent rose by approximately 39 per cent. This left previous tenants in an impossible financial position, unable to return to their previous homes. Lastly, as Inniss writes, at best misguided and at worse deliberately exclusionary public planning sought to gentrify communities through the promotion of mixed-income neighbourhoods (2007: 333). The impact of this policy has been to squeeze out the poor and, often, black: 'Desires to reduce the number of certain peoples in order to fight urban blight or crime stand as one of the most insurmountable barriers to the renewal of neighbourhoods chiefly occupied by blacks, and especially poor blacks' (2007: 353).

The implications of the failure to cultivate social textures in poorer suburbs post-disaster amount to the loss of community and identity for many; a deliberate domicide. When climate crises strike, a discriminatory politics, often racialised, of dwelling plays out in the Global North. Homeless and poor, African American victims of the hurricane were banished (Beckett and Herbert 2012).

RESISTING DOMICIDE

A politics of priority and expendability unfolds during times of climate crisis with 'marginal populations [experiencing] home loss as domicide through uneven preparedness, response and recovery in disaster policy' (Gorman-Murray et al. 2014: 243). With reference to preparedness, the examples of Queensland and New Orleans point to the discriminatory practices operating with regards to emergency warning and evacuation processes. As geographer Tim Cresswell (2008) notes, in the case of African American residents of New Orleans, the assumption of automobile ownership was detrimental and resulted in large numbers of stranded people when the levees broke. The assumption of car ownership was present in Queensland too, with refugee communities put at a disadvantage by this (MDA 2011). At an even more profound level, in the United States preparedness is impacted by the availability of safe and affordable housing. From the 1980s onwards, neoliberal policies 'emaciated federal support for low-income people . . . Housing assistance was particularly hard hit' (Beckett and Herbert 2012: 7). However, as Beckett and Herbert note, the withdrawal of federal funding to help citizens buy homes has not been evenly experienced. Indeed, 'for those who can afford to purchase a home, federal assistance is robust' (2012: 7).

The politics of expendability emerges the moment displacement occurs and the request for refuge or support is made. The process of dehumanising the victim of climate disasters is accelerated when this victim calls upon us to respond ethically to their abandonment. This is especially the case if the person comes from a community that has endured historic and ongoing racism or social disadvantage. An ethical orientation in this context requires precisely the same personal and social transformations I have outlined as essential to a relational notion of being-with-and-for-Others. We must, as a matter of urgency, undertake the same self-reflection regarding both how we dwell, and the authority of our own claims to dwelling. Part and parcel of this is taking up action in support of 'agency and resistance to domicide' (Nowicki 2014: 786), through the articulation of Other-oriented ethics in all disaster response, recovery and rebuilding frameworks. While HLP rights have so far been articulated in relation to geographical locations considered external to the Global North, I would suggest that the first step in ensuring a more ethical orientation in our disaster preparedness would be integrating the basic principles laid out in documents such as *The Pinheiro Principles* 2005.

CONCLUSION: GROUNDLESSNESS

We can ask here with Dan Bulley, 'why can we not just abandon the ontopological?' (2006: 650).[2] Bulley suggests that we cannot simply rid ourselves of the relationship between 'place' and 'identity'. This argument is supported when we appreciate the deep significance of place as a set of social and ecological 'attachments' in people's lives (Groves 2014; Nowicki 2014; Porteous and Smith 2001; Gorman-Murray et al. 2014). To abandon entirely the significance of these relationships, even at a conceptual level, would be a form of intellectual domicide.[3] As I have outlined, a deconstructive approach to dwelling takes us to the grounds of the home and reveals to us the dance we must engage in. This is a dance that at once understands what is going on when we regard dwelling as an act of possession at the same time as it is an opening onto the Other. It is this tension, the paradox of this double move, which provides the possibility for social and political change.

When deployed by international bodies such as the World Meteorological Organization, a discourse of 'crisis' is designed to incite responsibility: it is a call to action. Yet, in specific contexts, the language of climate crisis as a new norm has effects that extend beyond an urgent need to recognise and respond to the impacts of human life on the planet. A politics of crisis reverts to the logic of security in some instances, with the demarcation of boundaries and the withdrawal of social compassion. In each case outlined, vulnerability is unevenly distributed and responded to. Concepts such as resilience tend to be cast in individualist terms, failing to contextualise the structural and institutional factors which create or deny strong communities in context. Crisis demands from us the responsibility to take decisive action. Such action must be based on a negotiation of the political and economic conditions operating in variable contexts. Climate change, while a problem which introduces new issues in many regards, is still bound up in historic power relationships. This calls upon us to be attentive to the impacts not merely of the climate but also of the human structures which simultaneously create and sustain privilege and disadvantage.

CONCLUSION:
A FUTURE, OTHERWISE THAN THIS

'In the end, every society produces its migrants.'

Thomas Nail[1]

'This is tough: taking responsibility for something you can't see. But it's no tougher than taking responsibility for, say, not killing – you don't have to come up with a reason: you just do it and figure out why later.'

Timothy Morton[2]

'We cannot hide behind a wall', the former President of the United States, Barack Obama, declared to a cheering crowd in Berlin in May 2017 (Connolly 2017). In response to a question concerning the migration crises that currently beset Europe, Obama articulated what could be said was a humane but realist approach to the issue of human mobility. Retaining strict adherence to the dualistic thinking separating the citizen from the foreigner, Obama spoke of the tension of extending solidarity with victims in foreign lands while continuing to assert primary responsibility over the citizens and peoples within one's territorial borders, all the while cognisant of one's 'limited resources'. Political decisions have to continually negotiate a multiplicity of needs, and with reference to migration matters, the best approach, he argued, is to bolster foreign aid in order to 'create more opportunities for people in their own countries' (Obama cited in Connolly 2017).

It is true that we cannot hide behind a wall. Nevertheless, nor can we assume that foreign aid will sufficiently diminish the issue of human mobility across the globe. In the era of climate change, the very 'solutions' assumed to provide viable futures meet astounding limits. One cannot adapt to lands that are literally eroding into the oceans. Nor can one adapt sustainably in the face of frequent severe weather that tears apart communities and infrastructure over and again, providing little space for stability and relief in between. Collapsing climate adaptation into local development imperatives is only a partial political response to a problem that introduces a multiplicity of ethical dilemmas. Moreover, when Obama declares that a wall will not do, he is focused only on sites of human mobility coming towards the Global North from outside. However, as I have demonstrated, the issues of displacement, migration and human mobility are unfolding *within* Global North states too, and require sharp analysis of the racialised politics of hospitality and care at play. Whose dwelling is worthy of action? Whose is left to ruin? As I have been arguing in this book, climate funding must certainly support *in situ* development initiatives as a matter of urgent priority, but the evidence is coming in daily that we need to think beyond this. We must consider the profound ethical issues associated with both adaptation *and* human mobility in all its guises: from displacement, dispossession and immobility, to migration and communal relocation. These are taking place, in distinct ways, in both the Global South and North. Just as we cannot hide behind a wall, we cannot rest on a single solution.

At stake is our ethical imagination. On the one hand, our task is clear. In Timothy Morton's words, 'we're responsible for global warming. Formally responsible, whether or not we caused it, whether or not we can prove that we caused it. We're responsible for global warming simply because we are sentient. No more elaborate reason is required' (2010: 98). Shared sentience alone is enough to compel us to take responsibility. On the other hand, determining what to do in light of this primal responsibility is less straightforward. In this book I have put forward an ambitious philosophical and ethical intervention into the issues we face as a species in relation to climate change. Through a sustained philosophical inquiry into 'dwelling' I have made every effort to bring to the centre of our attention the seriousness of the challenge that confronts us. We are not charged with tinkering with the application of our social and political worlds. Our responsibility is much greater. Our critical awareness must reach into the very dynamics of possession, ownership and belonging we operate with, or what Heidegger calls our building, dwelling and thinking.

By drawing on Heidegger, this book has emphasised the need to combat the plight of dwelling we are currently enduring. We must resist the technologisation of the world (its reduction to mere resource) and instead turn onto the world with an orientation of care. More seriously still, taking inspiration from Levinas, I have illuminated the groundlessness of our 'dwelling'. What Levinas means by this is that what we consider to be stable and firm, like the home that has a door and closed windows, is inherently exposed to what is external to it. Our 'home' is part of, and opens onto, the world. Climate change confirms Levinas's point in a number of important ways. Most obviously, our dwelling place is threatened by the force of the changing climate. But at a more metaphysical level, our 'possession' of the dwelling is contested. We are called into being by the Other whom we are thereafter indebted to and responsible for. And moreover, within Derrida's deconstructive logic, it is the foreigner's question 'where?' that compels us to think otherwise the relationship between the self or sovereign state, place, placelessness and ethics. These philosophical questions take shape, and morph, in diverse political and geographical contexts.

Yet my discussion by no means exhausts the theoretical or practical ethical options available to us. I have offered one alternative to the more analytic focus of dominant paradigms, such as CBDR, that influence current climate policy. I believe the arguments presented in this book are well suited to doing the work of critique and transformation of both the 'self' who dwells in the world, and the model of sovereign statehood that, at present, defines the political and legal modalities of belonging to specific pieces of the earth itself. The task we are invited to participate in here is more radical than the normative appeals to distributive responsibility with its calculable, logical and rational deck of philosophical cards. However, even within the CBDR paradigm, when positioned in a way that can be perceived as radical, we can find ourselves called upon us to accept that 'asymmetrical impacts alone provide sufficient reason to require the rest of the world to give forcibly displaced EVs (extremely vulnerable peoples) the right to move into other countries' (Byravan and Rajan 2010: 251). From another perspective, Mathias Risse has utilised cosmopolitan justice frameworks to argue that 'humanity collectively owns the earth' and as such, nation-states such as Kiribati have a moral claim to the right to relocate elsewhere (2009: 283).

Ethics should be front and centre of our contemporary discourse on climate change. However, even if and when ethics takes its position in the spotlight, it should not be co-opted into doing the work of perpetuating patterns of political and economic privilege and disadvantage. Rather, inspired by a deconstructive approach, ethics should embrace the conditions of 'permanent destabilisation' that we find ourselves in (Sokoloff 2005: 343). The impermanence of all things, including institutions such as the nation-state, and the mortal self, is not the social fact of life to be approached with fear and loathing. Rather, what we should be afraid of are the implications of our political programmes driven by an impossible desire for permanency and exclusive security of these egoistic institutions at the expense of the Other. No, we cannot build a wall, but even if we do, we must recall that walls fall down, disintegrate, hide us, and imprison us within narrow, self-interested and paranoid social and political horizons. Thomas Nail is right to refer to the migrant as the political figure of our time; the figure that calls all political constructions into question. However, the migrant, the dispossessed, the displaced and so on are more than this. They are all figures that demand an ethical response. They demand from us a bigger imagination.

If we cannot face the challenge of negotiating hospitality with the human stranger, how will we navigate the myriad other dilemmas that climate change poses? If we cannot see a fellow member of the human species as worthy of dignity and hospitality, how will we deal with accepting the rights of the earth, as per the principles of Earth Jurisprudence?[3] In this book I have focused only on the ethical relationship between humans (and the attachments humans have to places, land and social textures), drawing attention to the metaphysical and ontological arguments that bind us to one another in ways that demand an ongoing responsiveness to the needs of others. But this focus is not because I am only interested in Other-oriented *human* rights. I made a pragmatic decision to focus on these responsibilities, but would like to suggest that the ideas put forward in this book invite us to dwell, that is *think* in the Heideggerian sense, on the implications of climate change on our responsibilities towards non-human Others, including the earth itself and all its sentient creatures. Dwelling on this earth surely requires that we begin to appreciate the sentience of all earthly things. Our responsibility is post-human in the sense that it encompasses all species. Moreover, the earth-as-dwelling will only continue to extend something resembling hospitality towards human beings so long as we begin to actually take care of it as our home, temporary as it may be! The earth has its own agency, something

Heidegger was attuned to, and a facet of thinking that has started to find its feet in new materialist literatures (see Bennett 2010; Coole and Frost 2010).

At stake in our relationship to the Other is the future of ourselves, the earth and all Others. For Derrida, it is only because of our capacity to remain open to the encounter with the Other, with 'the unforeseeable *itself*', that we can even speak of a future as such (1978: 118–19). The Other – the earth and all its visitors – are the reason why we can even imagine the future. Dwelling *in* and *on* the future of our social and ecological worlds gestures towards an ethical commitment to understanding the negotiation of political thought today as constructing the 'future' with the profound knowledge that there is always some risk inherent in this project. What is to come is only calculable to a point, only able to be forecasted up to a certain degree, and thus always excessive and unknown. Dwelling *in* the era of climate change also recognises the messy and non-linear nature of time itself; the past haunts, certainly, but for Derrida, so does the future. *The future belongs to ghosts*. The present, this moment now, is always subject to the contextual negotiation of the past which challenges, and the future which threatens and promises, extends beyond our reach yet is perpetually upon us. While the world has dealt with the possibility of apocalypse by human nuclear destruction, climate change is arguably the greatest threat we have faced as a species regarding our futurity and viability on this planet.

The crises that climate change presents, the decisions it requires, bring us back down to earth. We cannot take flight or retreat into denialism. If we turn to face the situation, we are given the opportunity to see with fresh eyes the inextricable nature of planetary and species survival and transformation. We are challenged with the task of imagining the future of all others and the earth itself. Is the future simply a continuation of what we have today? This is both absurd, and a disturbing possibility. The IPCC has made its scientific predictions clear: current rates of carbon emissions are set to result in dangerous climate change. The status quo, rather than lulling us into a dreamless sleep, should jolt us awake. By selecting terms that are up for grabs, my intention in this book has been to open debate on what constitutes dwelling, ethics and adaptation. By drawing on terms that are themselves subject to contention, my objective throughout this text has been to challenge the mainstream orientation we have towards climate change and human mobility. Finally, the aim of *Dwelling in the Age of Climate Change: The Ethics of Adaptation* has been to bring our attention to the plight of the marginalised and the poor. To dwell in and on the future is an invitation to enact an ethics of and for the Other.

NOTES

INTRODUCTION

1. Derrida (2006: 26).
2. The Royal Geographical Society (n.d.) defines the Global North–South divide very broadly as 'the concept of a gap between the Global North and the Global South in terms of development and wealth'. However, how this maps onto the global community is complex and subject to change. For instance, the Brandt Report released in the 1980s was able to draw a line between nation-states that could be considered privileged and those that were clearly disadvantaged. Today the situation we are dealing with means that inequalities are spread unevenly between, as well as within, countries. For a detailed discussion, see Wolvers et al. (2015). I will be using the terms to refer to relative privilege and capacity, both economic and political. Part and parcel of this usage is an awareness of the unequal global power relations at play. I will also switch between the use of Global North and wealthy West. Again, the point here is to highlight power relations. These are terms and categories that are open to critique and review; however, it is not the aim of this book to develop such arguments.
3. See the *Rio Declaration on Environment and Development* 1992 for an earlier definition of CBDR, particularly Principles 6 and 7, which outline the need to acknowledge the differing needs and capacities of 'developing' and 'developed' countries as well as the greater responsibility the 'developed' countries carry (see Honkonen 2009).
4. Again, the critique of sovereignty is transdisciplinary. From a cultural studies tradition, see Nick Mansfield's *The God Who Deconstructs Himself* (2010); in political science, see Jens Bartleson's *A Genealogy of Sovereignty* (1995); from philosophy, Ewa Ziarek's *An Ethics of Dissensus* (2001) is excellent; while Peter Gratton's *The State of Sovereignty* (2012) offers an overview of the conceptual history of sovereignty.
5. I use Other with the capitalisation in accordance with Levinas's work.

Levinas's capitalisation works to emphasise the importance and indeed the priority of the other person in the self–Other relationship.
6. See Derrida (1999) for an in-depth deconstruction of ethics/politics; see also Ziarek's (2001) analysis for an interpretation that recognises the operation of deconstruction in Levinas's texts. Fagan's *Ethics and Politics after Poststructuralism* asks: 'What do post-foundationalist accounts of ethics mean for practical politics?' (2013: 5). This debate is referenced again in Chapters 2 and 3.
7. We can understand dwelling to connote a set of practices specific to a culture and social group. However, I am coming at the question of dwelling from a philosophical perspective indebted especially to Heidegger, Levinas and Derrida.
8. For a discussion, see Derrida's 'Letter to a Japanese Friend' (1985). See also Gayatri Spivak's introduction to Derrida's *Of Grammatology* (1976).

CHAPTER 1

1. Arendt (1996: 110).
2. When 'man' is used in citations, I will retain the original word. However, I will not continue to include [sic] to indicate the sexist nature of this language. Please assume humankind in all instances where 'man' appears in quoted form.
3. 'Adjustment in natural or human systems in response to actual or expected climatic stimuli or their effects, which moderates harm or exploits beneficial opportunities. Various types of adaptation can be distinguished, including anticipatory and reactive adaptation, private and public adaptation, and autonomous and planned adaptation' (IPCC 2001: 882).
4. The tension between being-at-home and the disruption of this at-homeness will be developed in greater detail in Chapter 3. For now, we need to understand that the possibility of displacement is as empirically ancient to human beings as that of sedentariness.
5. In 2007, the Australian Greens Party introduced a private Bill for consideration in the Senate, the *Migration (Climate Refugees) Amendment Bill 2007*. The Bill was proposed as an amendment to Australia's immigration law contained in the *Migration Act 1958*. It included provisions for a class of visa for 'climate change refugees'.
6. Sometimes political calls to preserve roots deliberately cover over the actual relationship people may have to place. Rather than being enrooted or uprooted, vulnerable peoples may find themselves 'trapped in immobility' (Nassef 2014: 17). Immobility arises in conditions of political neglect, economic scarcity and political marginalisation. It is the experience of despair: of being restricted.

Notes

7. The appeal to 'blood and soil', for instance, is a recurrent thread in genocides (Kiernan 2007).
8. Andrew Vincent points out that welfare rights have a tradition in places like the US, with solidarity and communitarian movements (2010: 23). The denial or uptake of various rights, and their recognition in the first place, is a matter of politics. Jack Donnelly puts this well when he writes that 'Not every kind of systematic suffering leads to a recognised right. Politics largely determines whether any particular indignity, threat, or right is recognised' (2013: 98).
9. This is a complex issue with contested views on what ownership looks like. In the South Pacific, for instance, land rights are a social and political issue that requires the negotiation of universal frameworks with cultural traditions (see Corrin and Paterson 2007).

CHAPTER 2

1. Levinas (1969: 47).
2. Aminzadeh (2007: 243).
3. There are enormous tensions between Heidegger's and Levinas's philosophical and political works. I will not be able to address these here. There is also an extensive literature on these two philosophers. A good starting place is Chanter (2001).
4. In relation to adaptation we are confronted with the necessity of forms of 'modern' technology. For instance, we see the use of engineering and infrastructure technologies which are actually required to sustain community safety and durability. That is, rather than uprooting a community from its local dwelling place, technological solutions may enable people to remain-in-place. What is referred to as 'technology transfer' in adaptation literature (providing developing countries and marginalised communities with the knowledge, skills and capacity to develop the technological innovations required for its context) assists in building stronger local attachments and communities that may remain sustainable in the face of difficult and increasingly unpredictable climate conditions. Indeed, the right to technology has become a significant issue. It is reductive to dismiss modern technology as solely an instrument of destruction and uprootedness. Yet Irwin is right to note that 'technological framing alienates people from their local dwelling place' (2008: 32) in the sense that broader forces and capitalist imperatives have resulted time and again in the destruction of local community with large-scale development projects or the pressure to migrate for work in cities. It is this process that I refer to as 'technologisation'.
5. Levinas's *Totality and Infinity* (1969) is a thorough critique of Heidegger's

philosophy of Being and indeed, dwelling. Levinas regards Heidegger's work as a philosophy of power.

6. It is important to pause a moment and note the gendered nature of Levinas's conception of interiority and the originary welcome. For Levinas this is feminine. Levinas is aware of this problematic and the critique that not all homes are inhabited by a woman: 'Need one add that there is no question here of defying ridicule by maintaining the empirical truth or counter-truth that every home *in fact* presupposes a woman? The feminine has been encountered in this analysis as one of the cardinal points of the horizon in which the inner life takes place – and the empirical absence of the human being of "feminine sex" in a dwelling nowise affects the dimensions of femininity which remains open there, as the very welcome of the dwelling' (1969: 157–8). What is worth emphasising for the present analysis is the promise of welcome and safety (gentleness) found in the possibility of a private refuge. Again, this is a problematic assumption given the prevalence of domestic violence. I am not able to do justice to these issues in this book; see Nowicki for a discussion of how the home can be 'framed as a site of negativity, violence and repression' (2014: 788). This is an enormous debate in Levinasian scholarship on gender and dwelling that warrants attention. And yet, I must go on without further elaboration, understanding that in doing so there is the risk of essentialising femininity with passivity, gentleness and intimacy. See Chanter (2001) for a specific discussion. See Timothy Morton for an engaging ecological ethics that advocates for the 'feminine warmth that Levinas describes' (2010: 127–8).

CHAPTER 3

1. United Nations General Assembly (2016: 3).
2. For a discussion of the manner in which humanitarian discourses of migration have been co-opted into border security paradigms, see Vaughan-Williams (2015). Vaughan-Williams's focus is on the EU and he develops an argument which seeks to go beyond the 'rhetoric versus reality' debate that marks migration matters. He argues that remaining stuck in this binary has produced a *'crisis of humanitarian critique'* (2015: 4; original italics).
3. See Campbell (1998) for a critique of the sovereign state within the tradition of International Relations. See Sassen (1999) for a discussion of states and foreigners from a sociological position.
4. There has been a considerable upsurge of support for far-right political parties across Europe. In Austria, for instance, the Freedom Party went from winning 21 per cent of the popular vote in 2013 to 35 per cent in 2016, following the Syrian civil war, which witnessed mass refugee move-

ments across Europe. Denmark, France, the United Kingdom, Switzerland and Poland have also seen a rise in the popularity of anti-immigration political parties (Aisch et al. 2016). The Brexit vote in the United Kingdom in 2016 was driven by a strong anti-immigration stance. In Australia the 'Reclaim Australia' movement has gained some momentum and the re-election of Pauline Hanson and One Nation signals a swing, once more, to the far right. The United States has been plagued by racial violence for some time; most recently, the election of Donald Trump for President was won on the back of racist policies towards immigrants as well as a promise to return America to its insularity.

5. The notion of 'burden-sharing' emerged following WWII, in the 1950s. The 1951 *United Nations Convention Relating to the Status of Refugees* notes that the 'burden' of hosting refugees tends to be unevenly distributed and calls upon the international community of member states to co-operate and spread the load. More recently, the language has shifted to 'responsibility-sharing' rather than 'burden sharing'. The latter tends to name refugees as burdensome, while the former avoids this by emphasising the obligations of member states.

6. This is a contentious topic. Debate is ongoing with leading international legal scholars who either tend to reject the need for the extension of refugee rights to victims of climate change, or have sought to develop criteria for inclusion into refugee-like paradigms of protection (Biermann and Boas 2010). Moreover, NGO advocates for political asylum remain opposed to including climate refugees in the Refugee Convention. See Michelle Foster's brilliant book, *International Refugee Law and Socio-Economic Rights: Refuge from Deprivation* (2007).

7. My thanks go to the anonymous reviewer who provided this perspective in their feedback.

8. The issue of large-scale migrations has been particularly pressing following the Syrian refugee crisis and its impacts in Europe. Efforts to manage all forms of migration are under way within the United Nations. Former Secretary-General Ban Ki Moon posited that the international community can humanely respond to the reality of large migrant movements: 'With the necessary political will, the world's responses to large movements of people can be grounded in shared values of responsibility-sharing, non-discrimination and respect for human rights, while also taking full advantage of the opportunity migration provides to stimulate development and economic growth' (United Nations General Assembly 2016: 3).

9. 'Loss and Damages' emerged as early as the 1980s as an international discussion point for understanding the long-term, inevitable and largely irreversible impacts of climate change. These impacts are likely to hit the vulnerable and disadvantaged nation-states and communities

disproportionately (Warner and van der Geest 2013). In 2012, the Loss and Damages framework was consolidated by a Working Party at the Doha Climate Talks, only to be rejected by numerous influential nation-states in 2013 at the Warsaw Climate Talks. The reason for this rejection turns on the stronger emphasis the framework has on issues of accountability and responsibility and consequently on appropriate avenues for rehabilitation and compensation.

10. See Morss (2003) for a detailed critique of Douzinas. See Ziarek for a discussion of Wendy Brown's 'trenchant critique of liberal rights as a mode of subjection to state regulation' in relation to Levinasian Other-oriented rights (2001: 69).
11. For an argument in favour of open borders see Carens (1987, 2013). Carens's latest work, *The Ethics of Immigration* (2013), provides an interesting starting point, with a discussion that works within existing institutional structures (specifically the nation-state) before offering up a solid critique in the final chapters.
12. The relationship between metaphysics and ontology, and ethics and politics, is a major debate within Derridean scholarship across disciplines such as Cultural Studies, Political Science, International Relations and Philosophy. As I have been using these terms there are tensions that invariably emerge: on the one hand, I am describing and affirming the metaphysics Levinas advances (*a priori* ethics), while on the other, I am suggesting that at the moment of enactment this metaphysics meets the ontological. I do not seek to resolve this tension as my argument is that when we engage deconstructively we are dancing between and within possibilities without rest or end. In a parallel manner, ethical principles are in an ongoing conversation (or negotiation) with political decisions. For a range of perspectives on this debate, see Hägglund (2008); Critchley (1999); Fagan (2013); Mansfield (2013).
13. The work of the Nansen Initiative in developing a protection agenda has been vital. This organisation is endeavouring to foster international support for protection instruments related to environmental and climate displacement and migration.
14. This denial is not universal, but it does urge us to consider what we will do as a global community in the coming decades as the movement of people increases with the impacts of climate change. Some hope emerges when we notice that following the mass movement of refugees from Syria from 2011 onwards, Germany accepted in excess of 800,000 refugees, while Sweden welcomed the most in proportion to its population. However, at least in Sweden the rise of the far right political parties has accompanied this event.
15. Governance is another knotted and complex arena. The IOM is state-

funded and does not provide 'formal access for social justice organisation'; it is under no formal obligation to abide by human rights norms (Basok and Piper 2013: 261). See also Taran (2001) who observes a 'rapid' increase in the level of intergovernmental co-operation concerned with the management of migration without a commensurate emphasis on migrant rights.

CHAPTER 4

1. Ahmed (2013: 52).
2. In Chapter 1, I demonstrated that these rights can already be found in numerous international doctrines and amount to a 'protection perspective' demanding the recognition of basic rights to security and dignity in addition to those related to living, such as water (Leckie and Huggins 2011: 2). However, HLP brings these strands together under a core human rights paradigm.
3. A report commissioned by a number of prominent international research institutes and NGOs, titled, *Loss and Damage in Vulnerable Countries Initiative: Bangladesh Leading the Way on Loss and Damage* (Roberts 2012) provides an excellent summary of the emergence of Loss and Damages in international dialogues. Roberts provides an overview of the manner in which the framework has been taken up in relation to Bangladesh. See also Roberts and Pelling's review article, 'Climate change-related loss and damage: translating the global policy agenda for national policy processes' (2016). This issue reveals another layer to the ethical consequences and complications of climate change adaptation, which exceed the parameters of this book. For a useful starting point, see Page and Heyward (2017) who offer a normative compensatory justice argument regarding Loss and Damages, and McShane (2017) who unpacks the related but distinct issues of 'harm' and 'value' in relation to what constitutes Loss and Damages.
4. In 2013, when I interviewed CEO and founder, Muhammad Abu Musa, of the local Bangladeshi organisation, Association for Climate Refugees, I was provided with a summary sheet of the organisation's working terminology. Musa highlighted the international controversy surrounding the use of the term 'refugee', but insisted that the conditions of climate-induced migration could amount to such a degree of deprivation: 'Climate refugee refers to a person who has been forced by climate-induced disaster events to relocate to a place for permanent resettlement and does not have the scope to return to his or her habitual place of residence. Such a person has completely or partly lost his or her house and livelihood assets but has completely lost the land and has not been protected by the law

of the land for necessary compensation. Such a person has relocated not only individually but also along with other members of the family.'
5. BRAC is the largest non-governmental development in the world. It started in Bangladesh and is based there. It is a money-lending service that offers micro-credit to the poor.
6. Muhammad Abu Musa, personal communication, 27 July 2013.
7. Sajid Raiham, personal communication,18 July 2013.
8. Feldman and Geisler refuse the language of migration in this context, preferring instead to label it 'eviction': 'the lived experience of displacement occurs in stages and is not readily perceived as coercive, socially-induced eviction at any given moment' (2012: 980).
9. For an excellent analysis of how 'hope' is inscribed in, or denied to, different places, see Jansen and Löfving's edited collection *Struggles for Home* (2009).
10. The international organisation, Displacement Solutions, has analysed the challenges of relocation in relation to the need to acquire appropriate lands. They note that even though land acquisition programmes are vital to the provision of climate adaptation, 'what could have been a relatively straight forward land acquisition and land allocation process for Carteret Islanders has instead been a long drawn out complicated process' (2012: 30).
11. Md Shamsuddoha, personal communication, 24 July 2013.
12. Simon Critchley's extensive work on Levinas provides the most thorough and thoughtful consideration of ethics as interruption. Fagan critiques Critchley's position from a Derridean point of view.
13. See the earlier discussion related to the deconstruction of ethics and politics, metaphysics and ontology, in Chapter 2.
14. There is another issue at stake. Should the unconditional be regarded as always 'good'? See Mansfield (2013) for an excellent discussion of the relationship between unconditionality and violence.
15. Selective immigration policies were developed in the 1980s in the Organisation for Economic Co-operation and Development countries such as Australia and Canada. More than thirty years on, there remains an in-built inequity in the structures of economic migration. This inequality is further evident in the significant rise of skilled migrants leaving developing countries compared with that of low-skilled migrants.
16. See McAdam (2009) for an interesting discussion on the link and distinction between 'dignity' in a legal sense compared with a social, cultural and personal sense.
17. Nicholas Blake (2010) has written a useful piece on the need for equity and balance in our labour migration.

CHAPTER 5

1. In *Ghost Dance*.
2. My Warusam, Saibai man (Leung 2008).
3. John Altman and Kirrily Jordan provided a submission to the Garnaut Climate Change Review in 2008. This report documented the vulnerability of indigenous communities across Australia. The authors note that Australia is 'lagging' behind other nation-states in addressing the vulnerability of Indigenous peoples to climate change.
4. Relocation or adaptation measures for people will emerge as a more mainstream issue in Australia, especially once it starts to affect non-Indigenous Australians. For example, in the news report 'How climate change is affecting the wine we drink', we learn that the changing climate is affecting the grapes used to produce Australian wine (Huntley 2016).
5. Personal communication, 15 June 2013.
6. Personal communication, 16 June 2013.
7. In 2010, the Australian government prepared a document entitled the 'Priority Investment Communities – WA'. In this, it was concluded that 192 of the 287 remote communities in Western Australia were 'unsustainable' (O'Connor 2015).
8. The Sustainable Livelihoods Approach focuses on development activities that are participatory, people-centred, sustainable, partnered with public and private sectors, responsive to local conditions and operating at multiple levels (Serrat 2008).
9. The Australian Greens Party offers an ecological alternative to the major parties.

CHAPTER 6

1. Debord (1967: para. 4).
2. See Chapter 2.
3. For instance, cultural theorist Tom Cohen has urged us to 'disoccupy' the home and to open 'beyond the propriety of the *oikos*' (2012: 16–17). Can we disoccupy it? 'What might be disoccupied', Cohen says, 'would be the metaphorics [sic] of the home' (2012: 17). Cohen is tired of the manner in which themes of hospitality, the ethics of the Other (read Levinas), and sovereignty come to constitute what he regards as a loop or return to self-sovereignty when placed under the eye of the deconstructionist. His critical intervention reaches its scathing heights when he contends that 'The *aporia* of an era of climate change are structurally different from those that devolved on the torsions of Western metaphysics. They are not the aporia explored by Derrida around the figure of *hospitality*, taken

as endless refolding that keeps in place, while exposing, a perpetual and lingering logics that defers the *inhospitable*. (One mode of deconstruction as solicitation involves shaking the house or structure within which one finds oneself, and this circuit might itself be disturbed by a refusal to occupy)' (2012: 19). Cohen is rightly critical of the imperative to possess contained in the act of 'occupying'. Yet his reflections on what he regards as the limits of deconstruction do not take us outside of the dynamics of the home and of hospitality. Instead, this desire for an outside must be questioned for its implicit privilege. Only those who are safely housed, owners of sorts, can surely step outside. Only those who can, in a material sense, return to the presumed safety of the home can simultaneously make the call to abolish this entire metaphorical structure! If the home comes to constitute the metaphorical structure of the earth itself – as our 'home', our desired and desirable place of 'habitation', as the 'dwelling' place for all earthbound life – where do I go if I wish to abandon this, even symbolically or rhetorically? What sort of metaphorical structure comes to take its place?

CONCLUSION

1. Nail (2015: 15).
2. Morton (2010: 99).
3. For an introduction into this wonderful realm of analysis, see Peter Burdon's edited collection *Exploring Wild Law* (2011). See also Christopher Stone's classic *Should Trees Have Standing?* (2010). In this, Stone argues that trees should have rights of sort and that when rights are articulated in law, it 'brings into the legal system a flexibility and open-endedness that no series of specifically stated legal rules . . . can capture' (2010: 22).

BIBLIOGRAPHY

ABC (2011), 'Evacuees feel like refugees in their own city', Australian Broadcasting Corporation, 13 January, <http://www.abc.net.au/news/2011-01-13/evacuees-feel-like-refugees-in-their-own-city/1903442> (last accessed 7 December 2016).
Adger, N. W., Dessai, S., Goulden, M., Hulme, M., Lorenzoni, I., Nelson, D. R., Naess, L., Wolf, J. and Wreford, A. (2009), 'Are there social limits to adaptation to climate change?', *Climatic Change*, 93:3, 335–54.
Adger, N. W., Pulhin, J. M., Barnett, J., Dabelko, G. D., Hovelsrud, G. K., Levy, M., Oswald-Springer, U. and Vogel, C. H. (2014), 'Human security', in C. B. Field, V. R. Barros, D. J. Dokken, K. J. Mach, M. D. Mastrandrea, T. E. Bilir, M. Chatterjee, K. L. Ebi, Y. O. Estrada, R. C. Genova, B. Girma, E. S. Kissel, A. N. Levy, S. MacCracken, P. R. Mastrandrea and L. L. White (eds), *Climate Change 2014: Impacts, Adaptation, and Vulnerability Part A: Global and Sectoral Aspects. Contribution of Working Group II to the Fifth Assessment Report of the Intergovernmental Panel on Climate Change*, Cambridge and New York: Cambridge University Press, pp. 755–91.
Advisory Group on Climate Change and Human Mobility (2015), *Human Mobility in the Context of Climate Change*, UNFCCC-PARIS-COP-21, <http://www.unhcr.org/protection/environment/565b21bd9/human-mobility-context-climate-change-unfccc-paris-cop-21-recommendations.html> (last accessed 30 June 2016).
Afsar, R. (2003), 'Internal migration and the development nexus: the case of Bangladesh', in *Regional Conference on Migration, Development and Pro-Poor Policy Choices in Asia 2003*, Dhaka, Bangladesh: Refugee and Migratory Movements Research Unit, <http://r4d.dfid.gov.uk/PDF/Outputs/MigrationGlobPov/WP-CP2.pdf> (last accessed 23 December 2016).
Ahmed, N. (2013), 'Entangled earth', *Third Text*, 27:1, 44–53.
Ahmed, S., Castañeda, C. and Fortie, A. (2003), *Uprootings/Regroundings: Questions of Home and Migration*, Oxford: Berg Publishers.
Ahn, I. (2010), 'Economy of "invisible debt" and ethics of "radical hospitality":

toward a paradigm change of hospitality from "gift" to "forgiveness"', *Journal of Religious Ethics*, 38:2, 243–67.

Aisch, G., Pearce, A. and Rousseau, B. (2016), 'How far is Europe swinging to the Right?', *The New York Times*, 17 November, <http://www.nytimes.com/interactive/2016/05/22/world/europe/europe-right-wing-austria-hungary.html>(last accessed 24 November 2016).

Akter, T. (2009), *Climate Change and Flow of Environmental Displacement in Bangladesh*, Dhanmondi, Dhaka: Centre for Research and Action on Development.

Altman, J. C. and Jordan, K. (2008), *Impacts of Climate Change on Indigenous Australians: Submission to the Garnaut Climate Change Review*, Canberra: Centre for Aboriginal Economic Policy Research, Australian National University, <http://caepr.anu.edu.au/sites/default/files/Publications/topical/Altman_Jordan_Garnaut%20Review.pdf> (last accessed 23 December 2016).

American Declaration of the Rights and Duties of Man 1948, OAS, opened for signature April 1948, Bogotá, Colombia.

Aminzadeh, S. C. (2007), 'Moral imperative: the human rights implications of climate change', *Hastings International and Comparative Law Review*, 30:2, 231–67.

Anderson, S. (2014), 'Climate change intensifies risk of conflict, migration: IPCC report', *Special Broadcasting Service News*, 31 March, <http://www.sbs.com.au/news/article/2014/03/31/climate-change-intensifies-risk-conflict-migration-ipcc-report> (last accessed 7 December 2016).

Arendt, H. (1996), 'We refugees' (1943), in M. Robinson (ed.), *Altogether Elsewhere: Writers on Exile*, Boston and London: Faber & Faber, pp. 110–19.

Arendt, H. (1951), *The Origins of Totalitarianism*, New York: Schocken.

Arendt, H. (1958), *The Human Condition*, Chicago and London: University of Chicago Press.

Arendt, H. (2000), *The Portable Hannah Arendt*, London: Penguin Books.

Arendt, H. (2007), *Hannah Arendt: The Jewish Writings*, ed. J. Kohn. and R. Feldman, New York: Schocken.

Arthur, W. S. (2001), 'Indigenous autonomy in Australia: some concepts, issues and examples', Discussion Paper No. 220, Canberra: Centre for Aboriginal Economic Policy Research, Australian National University, <http://caepr.anu.edu.au/Publications/DP/2001DP220.php> (last accessed 23 December 2016).

Auchter, J. (2014), *The Politics of Haunting and Memory in International Relations*, Oxon and New York: Taylor & Francis.

Aulakh, R. (2013), 'Climate change forcing thousands in Bangladesh into slums of Dhaka', *thestar.com*, 16 February, <http://www.thestar.com/news/world/2013/02/16/climate_change_forcing_thousands_in_bangladesh_into_slums_of_dhaka.html> (last accessed 24 November 2016).

Bibliography

Australian Human Rights Commission (2008), *Native Title Report 2008*, Sydney: Australian Human Rights Commission, <https://www.humanrights.gov.au/our-work/aboriginal-and-torres-strait-islander-social-justice/publications/native-title-report-2008> (last accessed 23 December 2016).

Babbage, R. (1990), *The Strategic Significance of Torres Strait*, Canberra: Strategic and Defence Studies.

Baer, P. (2010), 'Adaptation to climate change: who pays whom?', in S. M. Gardiner, S. Caney, D. Jamieson and H. Shue (eds), *Climate Ethics: Essential Readings*, Oxford and New York: Oxford University Press, pp. 247–63.

Baker, G. (ed.) (2013), *Hospitality and World Politics*, Basingstoke: Palgrave Macmillan.

Banerjee, P. (2010), *Borders, Histories, Existences: Gender and Beyond*, London and New York: Sage.

Bangladesh Climate Change Resilience Fund (2016), 'Becoming a "Forest Saviour": community participation for conservation', <https://www.bccrf-bd.org/CaseStudies.html> (last accessed 28 November 2016).

Bardsley, D. K. and Hugo, G. J. (2010), 'Migration and climate change: examining thresholds of change to guide effective adaptation decision-making', *Population and Environment*, 32:2–3, 238–62.

Barkat, A., Zaman, S. and Raihan, S. (2000), *Khas Land: A Study on Existing Law and Practice*, Programme for Research on Poverty Alleviation, Grameen Trust, Grameen Bank, <http://www.hdrc-bd.com/admin_panel/images/notice/1380013777.03.%20khas%20land_%20a%20study%20on%20existing%20law%20and%20practice.pdf> (last accessed 6 December 2016).

Bartelson, J. (1995), *A Genealogy of Sovereignty*, Cambridge: Cambridge University Press.

Basok, T. and Piper, N. (2013), 'Justice for migrants: mobilising a rights based understanding of migration', in S. Ilcan (ed.), *Mobilities, Knowledge and Social Justice*, Montreal and London: McGill-Queen's University Press, pp. 255–76.

Bawden, T. (2013), 'Extreme heatwaves are predicted as the new normal for British summers by 2040', *Independent*, 15 August, <http://www.independent.co.uk/environment/climate-change/extreme-heatwaves-are-predicted-as-the-new-normal-for-british-summers-by-2040-8762336.html> (last accessed 23 December 2016).

Beckett, J. (1987), *Torres Strait Islanders: Custom and Colonisation*, Cambridge: Cambridge University Press.

Beckett, K. and Herbert, S. (2012), *Banished: The New Social Control in Urban America*, Oxford: Oxford Scholarship Online.

Behringer, W. (2010), *A Cultural History of Climate*, Cambridge and Malden, MA: Polity Press.

Beine, M., Docquier, F. and Rapoport, H. (2008), 'Brain drain and human

capital formation in developing countries: winners and losers', *Economic Journal*, 118:528, 631–52.
Bennett, J. (2010), *Vibrant Matter: A Political Ecology of Things*, Durham, NC and London: Duke University Press.
Bergmann, S. (2015), 'Sustainable development, climate change and religion', in E. Tomalin (ed.), *The Routledge Handbook of Religions and Global Development*, London: Routledge, pp. 389–404.
Betzold, C. (2015), 'Adapting to climate change in small island developing states', *Climatic Change*, 133:3, 481–9.
Biermann, F. and Boas, I. (2010), 'Preparing for a warmer world: towards a global governance system to protect climate refugees', *Global Environmental Politics*, 10:1, 60–88.
Biressi, A. (2004), '"Above the below": body trauma as spectacle in social/media space', *Journal of Cultural Research*, 8:3, 335–52.
Blake, N. (2010), 'Nursing migration: issues of equity and balance', *People and Place*, 18:2, 19–24.
Boas, I. (2015), *Climate Migration and Security: Securitisation as a Strategy in Climate Change Politics*, London and New York: Routledge.
Bose, P. S. (2016), 'Vulnerabilities and displacements: adaptation and mitigation to climate change as a new development mantra', *Area*, 48:2, 168–75.
Boulous Walker, M. (2017), *Slow Philosophy: Reading Against the Institution*, London: Bloomsbury.
Bronen, R. (2011), 'Climate-induced community relocations: creating an adaptive governance framework based in human rights doctrine', *New York University Review of Law and Social Change*, 35, 357–407.
Bronen, R. and Chapin, F. S. (2013), 'Adaptive governance and institutional strategies for climate-induced community relocations in Alaska', *PNAS: Proceedings of the National Academy of the United States of America*, 110:23, 9320–5.
Brown, G. W. (2010), 'The laws of hospitality, asylum seekers and cosmopolitan right: a Kantian response to Jacques Derrida', *European Journal of Political Theory*, 9:3, 308–17.
Brown O., Hammill, A. and McLeman, R. (2007), 'Climate change as the "new" security threat: implications for Africa', *International Affairs*, 83:6, 1141–54.
Brown, W. (1995), *States of Inquiry: Power and Freedom in Late Modernity*, Princeton: Princeton University Press.
Bulley, D. (2006), 'Negotiating ethics: Campbell, ontopology and hospitality', *Review of International Studies*, 32, 645–63.
Bulley, D. (2009), *Ethics as Foreign Policy: Britain, the EU and the Other*, Oxon and New York: Routledge.

Bulley, D. (2017), *Migration, Ethics and Power: Spaces of Hospitality in International Politics*, London: Sage.

Burdon, P. (ed.) (2011), *Exploring Wild Law: The Philosophy of Earth's Jurisprudence*, Kent Town: Wakefield Press.

Burggraeve, R. (2005), 'The good and its shadow: the view of Levinas on human rights as the surpassing of political rationality', *Human Rights Review* (January–March), 80–101.

Burggraeve, R. (2006), 'The Other and me: interpersonal and social responsibility in Emmanuel Levinas', *Revista Portuguesa de Filosofia* (April–December), 631–49.

Burroughs, W. J. (2005), *Climate Change in Prehistory: The End of the Reign of Chaos*, Cambridge and New York: Cambridge University Press.

Butler, J. (2004), *Precarious Life: The Powers of Mourning and Violence*, London and New York: Verso.

Butler, J. (2010), *Frames of War*, London and New York: Verso.

Butterly, L. (2013), 'Native title rights, regulations and licences: the Torres Strait sea claim', *The Conversation*, 8 August, <https://theconversation.com/profiles/lauren-butterly-100714/articles> (last accessed 7 December 2016).

Byravan, S. and Rajan, S. C. (2006), 'Providing new homes for climate change exiles', *Climate Policy*, 6, 247–52.

Byravan, S. and Rajan, S. C. (2010), 'The ethical implications of sea-level rise due to climate change', *Ethics & International Affairs*, 24:3, 239–60.

Campbell, D. (1998), *National Deconstruction: Violence, Identity, and Justice in Bosnia*, Minneapolis: University of Minnesota Press.

Cancun Agreements 2010, COP 16 UNFCCC, 11 December, Cancun, Mexico.

Caney, S. (2010), 'Climate change, human rights and moral thresholds', in S. M. Gardiner, S. Caney, D. Jamieson and H. Shue (eds), *Climate Ethics: Essential Readings*, Oxford and New York: Oxford University Press, pp. 163–80.

Carens, J. (1987), 'Aliens and citizens: the case for open borders', *The Review of Politics*, 49:2, 251–73.

Carens, J. (2013), *The Ethics of Immigration*, New York: Oxford University Press.

Carrick, D. (2011), *Climate Change: Indian Ocean*, audio podcast, Law Report, Radio National, ABC Radio, Sydney, 29 November, <http://www.abc.net.au/radionational/programs/lawreport/climate-change-indian-ocean/3695390> (last accessed 23 November 2016).

Carter, L. (2014), 'Criss-crossing highways: Pacific travelling and dwelling in times of global warming', *Journal of New Zealand & Pacific Studies*, 2:1, 57–67.

Carto, S. L., Weaver, A. J., Hetherington, R., Lam, Y. and Wiebe, E. C. (2009), 'Out of Africa and into an ice age: on the role of global climate change in the late Pleistocene migration of early modern humans out of Africa', *Journal of Human Evolution*, 56:2, 139–51.

Chanter, T. (2001), *Time, Death and the Feminine: Levinas with Heidegger*, Stanford: Stanford University Press.

Clark, N. (2011), *Inhuman Nature: Sociable Life on a Dynamic Planet*, London: Sage.

Cohen, H. (2012a), 'Relocation fears as ocean swallows Torres Strait Islands', *ABC News*, 3 March, <http://www.abc.net.au/news/2012-03-03/calls-for-seawalls-as-ocean-swallows-torres-strait-islands/3866564> (last accessed 23 December 2016).

Cohen, H. (2012b), *A Sinking Feeling in the Torres Strait*, audio podcast, Background Briefing, Radio National, ABC Radio, Sydney, 4 March, <http://www.abc.net.au/radionational/programs/backgroundbriefing/2012-03-04/3857272> (last accessed 7 December 2016).

Cohen, T. (2012), 'Murmurations – "climate change" and the defacement of theory', in T. Cohen (ed.), *Telemorphosis: Theory in the Era of Climate Change*, vol. 1, Michigan: Open Humanities Press, pp. 13–42.

Collerton, S. (2011), 'Torres Strait pleads for climate change action', *ABC Far North Queensland News*, 12 July, <http://www.abc.net.au/news/stories/2011/07/12/3267469.htm?site=farnorth> (last accessed 23 December 2016).

Connolly, K. (2017), 'Obama tells adoring crowd in Berlin: "We can't hide behind a wall"', *The Guardian*, 25 May, <https://www.theguardian.com/us-news/2017/may/25/barack-obama-draws-crowd-of-tens-of-thousands-in-berlin> (last accessed 20 October 2017).

Coole, D. and Frost, S. (2010), *New Materialisms: Ontology, Agency, and Politics*, Durham, NC and London: Duke University Press.

Cordes-Holland, O. (2008), 'The sinking of the Strait: the implications of climate change for Torres Strait Islanders' human rights protected by the ICCPR', *Melbourne Journal of International Law*, 9, 405.

Corrin, J. and Paterson, D. (2007), *Introduction to South Pacific Law*, Oxon and New York: Routledge-Cavendish.

Crary, J. (1989), 'Spectacle, attention and counter-memory', *October*, 50, 96–107.

Cresswell, T. (2008), 'Understanding mobility holistically: the case of Hurricane Katrina', in T. Sager and S. Bergmann (eds), *The Ethics of Mobilities: Rethinking Place, Exclusion, Freedom and Environment*, London and New York: Routledge, pp. 129–40.

Critchley, S. (1999), *The Ethics of Deconstruction: Derrida and Levinas*, Edinburgh: Edinburgh University Press.

Critchley, S. (2004), 'Five problems in Levinas's view of politics and the sketch of a solution to them', *Political Theory*, 32:2, 172–85.

Davis, C. (2005), 'Hauntology, spectres and phantoms', *French Studies*, 59:3, 373–9.

Dean, A. (2010), *Responses to Climate Change in the Torres Strait*, Honours thesis, University of Newcastle, New South Wales, Australia.
Dean, J. (2000), *Cultural Studies and Political Theory*, Ithaca and London: Cornell University Press.
Debord, G. (1967), *The Society of the Spectacle*, New York: Zone Books.
de Guchteneire, P. and Pécoud, A. (2006), 'International migration, border controls and human rights: assessing the relevance of a right to mobility', *Journal of Borderlands Studies*, 21:1, 69–86.
Dellink, R., den Elzen, M., Aiking, H., Bergsma, E., Berkhout, F., Dekker, T. and Gupta, J. (2009), 'Sharing the burden of financing adaptation to climate change', *Global Environmental Change*, 19, 411–21.
Department of Foreign Affairs and Trade (DFAT) (2016), *Overview of Australia's Assistance for Climate Change*, <http://dfat.gov.au/aid/topics/investment-priorities/building-resilience/climate-change/Pages/climate-change.aspx> (last accessed 14 September 2016).
Derrida, J. (1976), *Of Grammatology*, Baltimore: John Hopkins University Press.
Derrida, J. (1978), *Writing and Difference*, London and New York: Routledge.
Derrida, J. (1981), *Dissemination*, London and New York: Continuum.
Derrida, J. (1985), 'Letter to a Japanese Friend', in D. Wood and R. Bernasconi (eds), *Derrida and Différence*, Warwick: Parousia Press, pp. 1–5.
Derrida, J. (1999), *Adieu: To Emmanuel Levinas*, Stanford: Stanford University Press.
Derrida, J. (2000), 'Hostipitality', *Angelaki*, 5:3, 3–18.
Derrida, J. (2001a), *On Cosmopolitanism and Forgiveness*, London and New York: Routledge.
Derrida, J. (2001b), 'A discussion with Jacques Derrida', *Theory & Event*, 5:1, n.p.
Derrida, J. (2002), 'Hostipitality', in G. Anidjar (ed.), *Acts of Religion*, London and New York: Routledge, pp. 356–420.
Derrida, J. (2005), 'The principle of hospitality', *Parallax*, 11:1, 6–9.
Derrida, J. (2006), *Specters of Marx*, New York: Routledge.
Derrida, J. (2011), *Parages*, Stanford: Stanford University Press.
Displacement Solutions (2012), *Climate Displacement in Bangladesh: The Need for Urgent Housing, Land and Property (HLP) Solutions*, <http://www.displacementsolutions.org> (last accessed 27 December 2016).
Displacement Solutions (2013), *The Peninsula Principles on Climate Displacement within States*, <http://displacementsolutions.org/peninsula-principles/> (last accessed 27 December 2016).
Docker, E. (1970), *The Blackbirders: The Recruiting of South Seas Labour for Queensland, 1863–1907*, Sydney: Angus & Robertson.
Docquier, F. and Rapoport, H. (2009), 'Skilled migration: the perspective of developing countries', in J. Baghwati and G. Hanson (eds), *Skilled Migration:*

Prospects, Problems and Policies, New York: Russell Sage Foundation, pp. 247–85.

Doherty, B. (2012), 'Climate change castaways consider move to Australia', *The Sydney Morning Herald*, 7 February, <http://www.smh.com.au/environment/climate-change/climate-change-castaways-consider-move-to-australia-20120106-1pobf.html> (last accessed 7 December 20116).

Donnelly, J. (2013), *Universal Human Rights in Theory and Practice*, Ithaca, NY and London: Cornell University Press.

Douglas, J. (1900), *Past and Present of Thursday Island and Torres Strait*, Brisbane: The Outridge Painting Co. Printers and Publishers.

Drabinski, J. E. (2012), *Sensibility and Singularity: The Problem of Phenomenology in Levinas*, Albany: State University of New York Press.

Duce, S. J., Parnell, K. E., Smithers, S. G. and McNamara, K. E. (2010), *A Synthesis of Climate Change and Coastal Science to Support Adaptation in the Communities of the Torres Strait*, Cairns: Marine and Tropical Sciences Research Facility, Reef & Rainforest Research Centre.

Dufourmantelle, A. and Derrida, J. (2000), *Of Hospitality: Anne Dufourmantelle Invites Jacques Derrida to Respond*, Stanford: Stanford University Press.

Edkins, J. (2003), *Trauma and the Memory of Politics*, Cambridge: Cambridge University Press.

Edkins, J. (2006), 'Remembering relationality: trauma, time and politics', in D. Bell (ed.), *Memory, Trauma and World Politics*, Basingstoke: Palgrave Macmillan, pp. 99–115.

El-Hinnawi, E. (1985), *Environmental Refugees*, Nairobi: United Nations Environment Programme.

Fagan, M. (2013), *Ethics and Politics after Poststructuralism*, Edinburgh: Edinburgh University Press.

Faist, T. and Schade, J. (2013), *Disentangling Migration and Climate Change: Methodologies, Political Discourses and Human Rights*, Dordrecht: Springer.

Farbotko, C. (2010), 'Wishful sinking: disappearing islands, climate refugees and cosmopolitan experimentation', *Asia Pacific Viewpoint*, 51:1, 47–60.

Farran, S. (2007), *Human Rights in the South Pacific: Challenges and Changes*, Oxon and New York: Routledge-Cavendish.

Feldman, S. and Geisler, C. (2012), 'Land expropriation and displacement in Bangladesh', *The Journal of Peasant Studies*, 39:3–4, 971–93.

Feldman, S., Geisler, C. and Menon, G. A. (eds) (2011), *Accumulating Insecurity: Violence and Dispossession in the Making of Everyday Life*, Athens, GA and London: The University of Georgia Press.

Ferris, E. (2011), *The Politics of Protection: The Limits of Humanitarian Action*, Washington DC: Brookings Institution Press.

Foster, M. (2007), *International Refugee Law and Socio-Economic Rights: Refuge from Deprivation*, Cambridge: Cambridge University Press.

Friedlingstein, R. M., Rogelj, A. J., Peters, G. P., Canadell, J. G., Knutti, R., Luderer, G., Raupach, M. R., Schaeffer, M., van Vuuren, D. P. and Le Quéré, C. (2014), 'Persistent growth of CO_2 emissions and implications for reaching climate targets', *Nature Geoscience*, 7, 709–15.

Friese, H. (2004), 'Spaces of hospitality', trans. James Keye, *Angelaki*, 9:2, 67–79.

Friese, H. (2009), 'Limits of hospitality', *Paragraph*, 32:1, 51–68.

Frosh, P. and Pinchevski, A. (2009), 'Crisis-readiness and media witnessing', *The Communication Review*, 12:3, 295–304.

Gardiner, S. M., Caney, S., Jamieson, D. and Shue, H. (eds) (2010), *Climate Ethics: Essential Readings*, Oxford and New York: Oxford University Press.

Gardiner, S. and Hartzell-Nichols, L. (2012), 'Ethics and global climate change', *Nature Education Knowledge*, 3:10, 1–5.

Gardner, K. and Ahmed, Z. (2009), 'Degrees of separation: informal social protection, relatedness and migration in Biswanath, Bangladesh', *The Journal of Development Studies*, 45:1, 124–49.

Gelbspan, P. and Thea, F. G. V. (2013), *Land in the Struggle for Social Justice: Social Movement Strategies to Secure Human Rights*, Curitiba: Terra de Direitos.

Ghost Dance, film, directed by K. McMullen. United Kingdom: Channel Four Films, 1983.

Gibb, C. and Ford, J. (2012), 'Should the United Nations Framework Convention on Climate Change recognise climate migrants?', *Environmental Research Letters*, 7, 1–9.

Ginsberg, A. (1996), *Allen Ginsberg: Selected Poems 1947–1995*, London and New York: Penguin.

Giroux, H. A. (2006), 'Reading Hurricane Katrina: race, class, and the biopolitics of disposability', *College Literature*, 33:3, 171–96.

Giroux, H. A. (2008), *Against the Terror of Neoliberalism: Politics Beyond the Age of Greed*, Boulder, CO: Paradigm Publishers.

Gleeson, M. (2016), *Offshore: Behind the Wire on Manus and Nauru*, Sydney: Newsouth.

Glover, L. (2006), *Postmodern Climate Change*, London and New York: Routledge.

Goldenberg, S. (2013), 'America's climate refugees', *The Guardian*, 13 May, <http://www.theguardian.com/environment/interactive/2013/may/13/newtok-alaska-climate-change-refugees> (last accessed 2 November 2016).

Gorman-Murray, A., McKinnon, S. and Dominey-Howes, D. (2014), 'Queer domicide', *Home Cultures*, 11:2, 237–61.

Gourevitch, A. (2010), 'Environmentalism: long live the politics of fear', *Public Culture*, 22:3, 411–24.

Government of Kiribati (GoK) (2015), *Kiribati National Labour Migration Policy*, South Tarawa: Government of Kiribati.

Grant, H., Randerson, J. and Vidal, J. (2009), 'UK should open its borders to climate refugees, says Bangladeshi minister', *The Guardian*, 4 December, <http://www.guardian.co.uk/environment/2009/nov/30/rich-west-climate-change> (last accessed 24 November 2016).

Grasso, M. (2010), 'An ethical approach to climate adaptation finance', *Global Environmental Change*, 20, 74–81.

Gratton, P. (2012), *The State of Sovereignty: Lessons from the Political Fictions of Modernity*, New York: State University of New York Press.

Green, D. (2006), 'How might climate change affect island culture in the Torres Strait Islands?', CSIRO Marine and Atmospheric Paper 011, <http://web.science.unsw.edu.au/~donnag/docs/climateimpacts_TSIculture_report.pdf> (last accessed 27 December 2016).

Green, D., Jackson, S. and Morrison, J. (2009), *Risks from Climate Change to Indigenous Communities in the Tropical North of Australia*, Canberra: Department of Climate Change and Energy Efficiency.

Gregory, D. (2004), *The Colonial Present: Afghanistan, Palestine, Iraq*, Malden, MA and Oxford: Wiley-Blackwell.

Greisch, J. (2011), 'Being, the Other, the Stranger', in R. Kearney and K. Semonovitch (eds), *Phenomenologies of the Stranger: Between Hostility and Hospitality*, New York: Fordham University Press, pp. 215–31.

Grossberg, L. (2006), 'Does cultural studies have futures? Should it? (Or what's the matter with New York?)', *Cultural Studies, Contexts and Conjunctures*, 1, 1–32.

Groves, C. (2014), *Care, Uncertainty and Intergenerational Ethics*, Basingstoke: Palgrave Macmillan.

Hägglund, M. (2008), *Radical Atheism: Derrida and the Time of Life*, Stanford: Stanford University Press.

Hall, S. (1990), 'The emergence of cultural studies and the crisis of the humanities', *October*, 53, 11–23.

Hammer, C. (2012), *The Coast*, Carlton: Melbourne University Press.

Hand, S. (ed.) (1989), *The Levinas Reader: Emmanuel Levinas*, Malden, MA and Oxford: Wiley-Blackwell.

Hannam, P. (2014), 'As climate changes, world weather agency calls for new baseline', *Sydney Morning Herald*, 10 July, <http://www.smh.com.au/environment/climate-change/as-climate-changes-world-weather-agency-calls-for-new-baseline-20140710-zt34x.html> (last accessed 23 December 2016).

Head, L. (2010), 'Cultural ecology: adaptation – retrofitting a concept', *Progress in Human Geography*, 34:2, 234–42.

Heidegger, M. (1978a), 'Building dwelling thinking', in D. F. Krell (ed.), *Martin Heidegger: Basic Writings*, London and Henley: Routledge and Kegan Paul, pp. 318–40.

Heidegger, M. (1978b), 'The question concerning technology', in D. F. Krell (ed.), *Martin Heidegger: Basic Writings*, London and Henley: Routledge and Kegan Paul, pp. 287–317.

Heidegger, M. (2008), *Being and Time*, New York: Harper Perennial.

Hillmann, F., Pahl, M., Rafflenbeul, B. and Sterly, H. (eds) (2015), *Environmental Change, Adaptation and Migration: Bringing in the Region*, London and New York: Palgrave Macmillan.

Hodes, J. (1998), *Migration to Cairns before WWII*, Master's thesis, Central Queensland University, Rockhampton, Australia.

Hogeveen, B. and Freistadt, J. (2013), 'Hospitality and the homeless: Jacques Derrida in the neoliberal city', *Journal of Theoretical and Philosophical Criminology*, 5:1, 39–63.

Honkonen, T. (2009), 'The principle of common but differentiated responsibility in post-2012 climate negotiations', *Reciel*, 18:3, 257–67.

Howells, M., O'Brien, C. and Sexton-McGrath, K. (2011), 'Bligh hears push for Torres Strait self-determination', *ABC News*, 29 August, <http://www.abc.net.au/news/2011-08-29/bligh-hears-push-for-torres-strait-self-determination/2859688> (last accessed 23 December 2016).

Huntley, R. (2016), 'How climate change is affecting the wine we drink', *Future Tense*, Australian Broadcasting Authority, 4 December, <http://www.abc.net.au/news/2016-12-04/how-climate-change-is-affecting-the-wine-we-drink/8074252> (last accessed 14 December 2016).

Hurst, D. (2011), 'Three-quarters of Queensland a disaster zone', *Brisbane Times*, 11 January, <http://www.brisbanetimes.com.au/environment/weather/threequarters-of-queensland-a-disaster-zone-20110111-19mf8.html> (last accessed 23 December 2016).

Huysmans, J. (2006), *The Politics of Insecurity: Fear, Migration and Asylum in the EU*, London and New York: Routledge.

Hyndman, J. (2000), *Managing Displacement: Refugees and the Politics of Humanitarianism*, Minneapolis: University of Minnesota Press.

Inniss, L. B. (2007), 'A domestic right of return?: Race, rights, and residency in New Orleans in the aftermath of Hurricane Katrina', *Boston College Third World Law Journal*, 27:2, 325–73.

Inter-Agency Standing Committee (2011), *IASC Operational Guidelines on the Protection of Persons in Situations of Natural Disaster*, Washington DC: The Brookings–Bern Project on Internal Displacement, <http://www.ohchr.org/Documents/Issues/IDPersons/OperationalGuidelines_IDP.pdf> (last accessed 27 December 2016).

Intergovernmental Panel on Climate Change (IPCC) (2001), *Climate Change 2001: Impacts, Adaptation and Vulnerability*, Cambridge: Cambridge University Press.

Intergovernmental Panel on Climate Change (IPCC) (2007), *Contribution of Working Groups I, II and III to the Fourth Assessment Report of the*

Intergovernmental Panel on Climate Change, Cambridge: Cambridge University Press.

Intergovernmental Panel on Climate Change (IPCC) (2012), 'Summary for policymakers', in C. B. Field, V. Barros, T. F. Stocker, D. Qin, D. J. Dokken, K. L. Ebi, M. D. Mastrandrea, K. J. Mach, G.-K. Plattner, S. K. Allen, M. Tignor and P. M. Midgley (eds), *Managing the Risks of Extreme Events and Disasters to Advance Climate Change Adaptation*, Special Report of Working Groups I and II of the IPCC, Cambridge and New York: Cambridge University Press, pp. 3–21.

Intergovernmental Panel on Climate Change (IPCC) (2013), 'Summary for policymakers', in T. F. Stocker, D. Qin, G. K. Plattner, M. Tignor, S. K. Allen, J. Boschung, A. Nauels, Y. Xia, V. Bex and P. M. Midgley (eds), *Climate Change 2013: The Physical Science Basis*, Contribution of Working Group I to the Fifth Assessment Report of the Intergovernmental Panel on Climate Change, Cambridge: Cambridge University Press, pp. 3–29.

Intergovernmental Panel on Climate Change (IPCC) (2014), 'Climate Change 2014: Synthesis Report Summary for Policymakers', <https://www.ipcc.ch/pdf/assessment-report/ar5/syr/AR5_SYR_FINAL_SPM.pdf> (last accessed 28 December 2016).

International Convention on the Protection of the Rights of All Migrant Workers and Members of Their Families 2003, opened for signature 18 December 1990, entered into force 1 July 2003.

International Covenant on Civil and Political Rights 1966, opened for signature by General Assembly resolution 2200A (XXI) of 16 December 1966, entry into force 23 March 1976, New York.

International Covenant on Economic, Social and Cultural Rights 1966, adopted and opened for signature, ratification and accession by General Assembly resolution 2200A (XXI) of 16 December 1966.

International Organisation for Migration (IOM) (2015a), *Global Migration Trends 2015 Factsheet*, <http://iomgmdac.org/global-trends-2015-factsheet/> (last accessed 6 September 2016).

International Organisation for Migration (IOM) (2015b), *Migration and Climate Change*, <https://www.iom.int/migration-and-climate-change-0> (last accessed 6 September 2016).

International Organisation for Migration (IOM) (2015c), *Migration in a World in Disarray: IOM Director General*, <https://www.iom.int/news/migration-world-disarray-iom-director-general> (last accessed 6 September 2016).

Irwin, R. (2008), *Heidegger, Politics, and Climate Change*, London: Continuum.

Isaac, J. C. (1996), 'A new guarantee on earth: Hannah Arendt on human dignity and the politics of human rights', *The American Political Science Review*, 90:1, 61–73.

Islam, R. (2013), 'Capital faces deluge of climate refugees', *Dhaka Courier*,

31 January, <http://www.dhakacourier.com.bd/?p=9952> (last accessed 24 November 2016).
Jaggar, A. (2013), 'Feminist ethics', in H. LaFollette and I. Persson (eds), *The Blackwell Guide to Ethical Theory*, Malden, MA and Oxford: Wiley-Blackwell, pp. 433–60.
Jamison, D. (2007), 'The moral and political challenges of climate change', in S. C. Moser and L. Dilling (eds), *Creating a Climate for Change: Communicating Change and Facilitating Social Change*, Cambridge: Cambridge University Press, pp. 475–85.
Jansen, S. and Löfving, S. (eds) (2009), *Struggles for Home: Violence, Hope and the Movement of People*, New York and Oxford: Berghahn Books.
Jones, V. (2005), 'Black people "loot" food . . . White people "find" food', *Huffington Post*, 1 September, <http://www.huffingtonpost.com/van-jones/black-people-loot-food-wh_b_6614.html> (last accessed 1 June 2017).
Joronen, M. (2011), 'Dwelling in the sites of finitude: resisting the violence of the metaphysical globe', *Antipode*, 43:4, 1127–54.
Kälin, W. and Schrepfer, N. (2012), *Protecting People Crossing Borders in the Context of Climate Change: Normative Gaps and Possible Approaches*, Legal and Protection Policy Research Series, Geneva: United Nations High Commissioner for Refugees.
Karasapan, O. (2015), 'Refugees: displaced from the Paris climate change agreement?', Brookings Institution, 7 December, <https://www.brookings.edu/blog/future-development/2015/12/07/refugees-displaced-from-the-paris-climate-change-agreement/> (last accessed 16 September 2016).
Kartiki, K. (2011), 'Climate change and migration: a case study from rural Bangladesh', *Gender and Development*, 19:1, 23–38.
Kavka, M. (2013), 'Postmodern Jewish ethical theories', in E. N. Dorff and J. K. Crane (eds), *The Oxford Handbook of Jewish Ethics and Morality*, Oxford: Oxford University Press, pp. 287–301.
Kiernan, B. (2007), *Blood and Soil: A World History of Genocide and Extermination from Sparta to Darfur*, New Haven, CT: Yale University Press.
Klein, N. (2016), 'Let them drown: the violence of othering in a warming world', *London Review of Books*, 38:11, 11–14.
Koser, K. (2012), *Environmental Change and Migration: Implications for Australia*, Sydney: Lowy Institute for International Policy.
Kythreotis, A. (2012), 'Progress in global climate change politics?', *Progress in Human Geography*, 36:4, 457–74.
Lacy Swing, W. (2015), 'Managing climate driven migration', in *Climate2020: Facing the Future*, London: United Nations Association, <http://e59114bec18f33b2ba6d-67d853478b97815e7adb8b9373d7dc7d.r53.cf2.rackcdn.com/CLIMATE2020.pdf> (last accessed 27 December 2016), pp. 77–80.

Laczko, F. and Aghazarm, C. (eds) (2009), *Migration, Environment and Climate Change: Assessing the Evidence*, Geneva: International Organisation of Migration.

Law, J. (2004), *After Method: Mess in Social Science Research*, London and New York: Routledge.

Leber, R. (2012), 'As "Frankenstorm" barrels toward East Coast, newspaper coverage ignores connection to climate change', *Think Progress Blog*, weblog, <http://thinkprogress.org/climate/2012/10/26/1097761/frankenstorm-sandy-climate/> (last accessed 23 December 2016).

Leckie, S. (2005), *HLP Rights in Post-Conflict Societies*, Geneva: Department of International Protection, United Nations High Commissioner for Refugees.

Leckie, S. and Huggins, C. (2011), *Conflict and Housing, Land and Property Rights: A Handbook on Issues, Frameworks and Solutions*, Cambridge: Cambridge University Press.

Leckie, S. and Lewis, D. (2010), 'Kiribati and Tuvalu will drown without global climate action', *The Ecologist Blog*, weblog, 11 November, <http://www.theecologist.org/blogs_and_comments/commentators/other_comments/680886/kiribati_and_tuvalu_will_drown_without_global_climate_action.html> (last accessed 23 December 2016).

Leung, C. (2008), 'Unlocking the memories of islands' tides of change', *The Age*, 21 March, <http://www.theage.com.au/news/environment/unlocking-the-memories-of-islands-tides-of-change/2008/03/20/1205602581720.html> (last accessed 23 December 2016).

Levin, S. (2010), 'Climate change: not all black and white', *ABC News Environment Feature*, 28 June, <http://www.abc.net.au/environment/articles/2010/06/28/2939097.htm> (last accessed 23 December 2016).

Levinas, E. (1969), *Totality and Infinity: An Essay on Exteriority*, Pittsburgh: Duquesne University Press.

Levinas, E. (1981), *Otherwise Than Being or Beyond Essence*, Pittsburgh: Duquesne University Press.

Levinas, E. (1998), *Entre Nous: On Thinking-of-the Other*, New York: Columbia University Press.

Levy, M. (2011), 'Levy to pay for $5.6b flood bill', *Sydney Morning Herald*, 27 January, <http://www.smh.com.au/business/levy-to-pay-for-56b-flood-bill-20110127-1a64x.html> (last accessed 23 December 2016).

Locke, J. T. (2009), 'Climate change-induced migration in the Pacific region: sudden crisis and long-term developments', *The Geographical Journal*, 175:3, 171–80.

Loughry, M. and McAdam, J. (2008), 'Kiribati: relocation and adaptation', *Forced Migration Review*, 31, 51–2.

McAdam, J. (2009), 'International refugee law and socio-economic rights: refuge from deprivation', *Melbourne Journal of International Law*, 10:2, 579–95.

McAdam, J. (2011), 'Swimming against the tide: why a climate change displacement treaty is not the answer', *International Journal of International Law*, 22:1, 2–27.

McAdam, J. (2013), 'Caught between homelands', *Inside Story*, 15 March, <http://inside.org.au/caught-between-homelands/> (last accessed 23 December 2016).

McAdam, J. and Saul, B. (2010), 'Displacement with dignity: international law and policy responses to climate change migration and security in Bangladesh', *German Yearbook of International Law*, 53, 233–89.

McClain, L. (1995), 'Inviolability and privacy: the castle, the sanctuary, and the body', *Yale Journal of Law and the Humanities*, 7:1, 195–241.

McNamara, K. E. (2015), 'Cross-border migration with dignity in Kiribati', *Forced Migration Review*, 49, 62.

McNamara, K. E. and Gibson, C. (2009), '"We don't want to leave our land": Pacific ambassadors at the United Nations resist the category of "climate refugees"', *Geoforum*, 40:3, 475–83.

McNamara, K. E., Smithers, S. G., Westoby, R. and Parnell, K. (2012), *Limits to Climate Change Adaptation for Low-Lying Communities in the Torres Strait*, Gold Coast: National Climate Change Adaptation Research Facility.

McQuillan, M. (2012), *Deconstruction without Derrida*, London: Bloomsbury.

McShane, K. (2017), 'Values and harms in loss and damage', *Ethics, Policy & Environment*, 20:2, 129–42.

Maldonado, J. K., Shearer, C., Bronen, R., Peterson, K. and Lazrus, H. (2013), 'The impact of climate change on tribal communities in the US: displacement, relocation, and human rights', *Climatic Change*, 120:3, 601–14.

Malkki, L. (1992), 'National geographic: the rooting of peoples in the territorialisation of national identity among scholars and refugees', *Cultural Anthropology*, 7:1, 24–44.

Manning, P. (2005), *Migration in World History*, Oxon and New York: Routledge.

Mansfield, N. (2008), '"There is a spectre haunting . . .": ghosts, their bodies, some philosophers, a novel and the cultural politics of climate change', *borderlands ejournal*, 7:1, <http://www.borderlands.net.au/vol7no1_2008/mansfield_climate.htm> (last accessed 27 December 2016).

Mansfield, N. (2010), *The God Who Deconstructs Himself: Sovereignty and Subjectivity between Freud, Bataille and Derrida*, New York: Fordham University Press.

Mansfield, N. (2012), '"There must be decision": climate change justice', *Australian Humanities Review*, 52, <http://www.australianhumanitiesreview.org/archive/Issue-May-2012/mansfield.html> (last accessed 26 October 2016).

Mansfield, N. (2013), 'Derrida, sovereignty and violence', *Lo Sguardo – Rivista di Filosofia*, 13, 143–62.
Martin, L. (2011), 'Disaster funds to target QLD's homeless', *Sydney Morning Herald*, 8 March, <http://news.smh.com.au/breaking-news-national/disaster-funds-to-target-qlds-homeless-20110308-1blue.html> (last accessed 23 December 2016).
Martin, M., Billah, M., Siddiqui, T., Black, R. and Kniveton, D. (2013), 'Policy analysis: climate change and migration Bangladesh', working paper 4, Refugee and Migratory Movements Research Unit, University of Dhaka, and Sussex Centre for Migration Research, University of Sussex, <http://migratingoutofpoverty.dfid.gov.uk/files/file.php?name=wp4-ccrm-b-policy.pdf&site=354> (last accessed 27 December 2016).
Martin, S. F. (2010), 'Climate change, migration and governance', *Global Governance*, 16:3, 397–414.
Martin, S. F. (2012), 'Environmental change and migration: legal and political frameworks', *Environment and Planning C: Government and Policy*, 30:6, 1045–60.
Martin, S. (2014), 'Taking stock of human mobility in initial National Adaptation Programmes of action and plans', in K. Warner, W. Kälin, S. Martin, Y. Nassef, S. Lee, S. Melde, H. Entwisle Chapuisat, M. Franck and T. Afifi (eds), *Integrating Human Mobility Issues within National Adaptation Plans*, UN University – Nansen Initiative Joint Policy Brief #2, Tokyo: United Nations University, pp. 18–22.
Martin, S. F., Weerasinghe, S. and Taylor, A. (eds) (2014), *Migration and Humanitarian Crises: Causes, Consequences and Responses*, New York and London: Routledge.
Masquelier, A. (2006), 'Why Katrina's victims aren't refugees: musings on a "dirty" word', *American Anthropologist*, 108:4, 735–43.
Milman, O. (2015), 'UN drops plan to help move climate-change affected people', *The Guardian*, 7 October, <https://www.theguardian.com/environment/2015/oct/07/un-drops-plan-to-create-group-to-relocate-climate-change-affected-people> (last accessed 5 September 2016).
Ministry of Environment (2010), *National Adaptation Programme of Action (NAPA) to Climate Change*, Kathmandu: Government of Nepal.
Ministry of the Environment and Forests (2008), *Bangladesh Climate Change Strategy and Action Plan 2008*, Dhaka: Government of the People's Republic of Bangladesh.
Ministry of the Environment and Forests (2009), *Bangladesh Climate Change Strategy and Action Plan 2009*, Dhaka: Government of the People's Republic of Bangladesh.
Ministry of Environment and Natural Resources (2016), *Kenya National Adaptation Plan 2015–2030*, Nairobi: Government of Republic of Kenya.

Morsheed, M. (2007), *Indigenous Coping Mechanisms in Combating Flood*, MA thesis, BRAC University, Dhaka, Bangladesh.

Morss, J. R. (2003), 'Saving human rights from its friends: a critique of the imaginary justice of Costas Douzinas', *Melbourne University Law Review*, 27, 889–904.

Morton, T. (2010), *The Ecological Thought*, Cambridge, MA and London: Harvard University Press.

Mountz, A. (2011), 'The enforcement archipelago: detention, haunting, and asylum on islands', *Political Geography*, 30:3, 118–28.

Müller, B., Höhne, N. and Ellermann, C. (2009), 'Differentiating (historic) responsibilities for climate change', *Climate Policy*, 9:6, 593–611.

Müller, C., Waha, K., Bondeau, A. and Heinke, J. (2014), 'Hotspots of climate change impacts in sub-Saharan Africa and implications for adaptation and development', *Global Change Biology*, 20:8, 2505–17.

Mulrennan, M. E. and Scott, C. H. (2000), 'Mare Nullius: indigenous rights in saltwater environments', *Development and Change*, 32, 681–708.

Mulrennan, M. and Scott, C. (2001), 'Indigenous rights and control of the sea in the Torres Strait', *Indigenous Law Bulletin 2*, 5:5.

Multicultural Development Association (MDA) (2011), *Queensland Floods: Commission of Inquiry 2011: Submission by the MDA*, Brisbane: Multicultural Development Association.

Murdoch, L. (2011), 'Mopping up brings relief for returnees', *The Age*, 15 January, <http://www.theage.com.au/environment/weather/mopping-up-brings-relief-for-returnees-20110114-19rck.html> (last accessed 23 December 2016).

Murray, W. E. and Overton, J. D. (2011), 'Neoliberalism is dead, long live neoliberalism? Neostructuralism and the international aid regime of the 2000s', *Progress in Development Studies*, 11:4, 307–19.

Myers, N. (1993), *Ultimate Security: The Environmental Basis of Political Instability*, New York and London: W. W. Norton.

Nail, T. (2015), *The Figure of the Migrant*, Stanford: Stanford University Press.

Nakata, M. (2007), *Disciplining the Savages – Savaging the Disciplines*, Canberra: Aboriginal Studies Press.

Nandy, G. and Mehedi, H. (2010), *Climate Change: Voices of the Victims from Bangladesh Coast*, Coastal Campaign Group: Campaign for Sustainable Rural Livelihoods, Khulnah: humanitywatch and Uttaran, <https://issuu.com/humanitywatch/docs/climate_migration_aila> (last accessed 27 December 2016).

Nansen Initiative (2015), *Agenda for the Protection of Cross-Border Displaced Persons in the Context of Disasters and climate Change, Volume II*, <https://www.nanseninitiative.org/wp-content/uploads/2015/02/PROTECTION-AGENDA-VOLUME-2.pdf> (last accessed 6 September 2016).

Nassef, Y. (2014), 'National adaptation plans and building adaptive capacity', in K. Warner, W. Kälin, S. Martin, Y. Nassef, S. Lee, S. Melde, H. Entwisle Chapuisat, M. Franck and T. Afifi (eds), *Integrating Human Mobility Issues within National Adaptation Plans*, UN University – Nansen Initiative Joint Policy Brief #2, Tokyo: United Nations University, pp. 16–17.

Nelson, C. and Gaonkar, D. P. (1996), *Disciplinarity and Dissent in Cultural Studies*, London and New York: Routledge.

ninemsn (2011), 'PM plays with kids at evacuation centre', *ninemsn*, 12 January, <http://news.ninemsn.com.au/national/floods/8195723/gillard-meeting-brisbane-flood-evacuees> (last accessed 23 December 2016).

Nowicki, M. (2014), 'Rethinking domicide: towards an expanded critical geography of home', *Geography Compass*, 8:11, 785–95.

O'Collins, M. (1990), 'Carteret Islanders at the atolls resettlement scheme: a response to land loss and population growth', in J. C. Pernetta and P. J. Hughes (eds), *Implications of Expected Climate Changes in the South Pacific Region: An Overview*, Regional Seas Reports and Studies No. 128, Nairobi: United Nations Environment Programme, pp. 247–69.

O'Connor, A. (2015), 'Leaked document reveals 192 WA Aboriginal communities deemed unsustainable in 2010', *ABC News*, 25 March, <http://www.abc.net.au/news/2015-03-24/federal-review-reveals-192-communities-deemed-unsustainable/6343570> (last accessed 14 December 2016).

Okereke, C. (2010), 'Climate justice and the international regime', *Wiley Interdisciplinary Reviews: Climate Change*, 1:3, 462–74.

O'Neill, C., Green, D. and Lui, W. (2012), 'How to make climate change research relevant for Indigenous communities in Torres Strait, Australia', *Local Environment: The International Journal of Justice and Sustainability*, 17:10, 1104–20.

Osborne, E. (2009), *Throwing off the Cloak: Reclaiming Self-Reliance in the Torres Strait*, Canberra: Aboriginal Studies Press.

Pacific-Australia Climate Change Science and Adaptation Planning Program (2016), *Current and Future Climate of Tuvalu*, brochure, <http://www.pacificclimatechangescience.org/wp-content/uploads/2013/06/4_PACCSAP-Tuvalu-10pp_WEB.pdf> (last accessed 4 October 2016).

Page, E. A. and Heyward, C. (2017), 'Compensating for climate change loss and damage', *Political Studies*, 65:2, 356–72.

Parr, A. (2012), *The Wrath of Capital*, New York: Columbia University Press.

Pécoud, A. and de Guchteneire, P. (2006), 'International migration, border controls and human rights: assessing the relevance of a right to mobility', *Journal of Borderlands Studies*, 21:1, 69–86.

Pécoud, A. and de Guchteneire, P. (2007), *Migration without Borders: Essays on the Free Movement of People*, New York: Berghahn Books.

Piguet, E., Pécoud, A. and de Guchteneire, P. (2011), 'Migration and climate change: an overview', *Refugee Survey Quarterly*, 30:3, 1–23.

The Pinheiro Principles (2005), United Nations Principles on Housing and Property Restitution for Refugees and Displaced Persons, Geneva: Centre on Housing Rights and Evictions.

PMSEIC Independent Working Group (2007), *Climate Change in Australia: Regional Impacts and Adaptation – Managing the Risk for Australia*, report prepared for the Prime Minister's Science, Engineering and Innovation Council, Canberra, Canberra: Government of Australia.

Poi Poi, R., Basiu, C., Williams, R., Sagankaz, J. and Yusia, J. (2000), *Saibai to Bamaga: The Migration from Saibai to Bamaga on the Cape York Peninsula*, brochure, Bamaga, Queensland: Bamaga Island Council.

Poleshchuk, I. (2010), 'Heidegger and Levinas: metaphysics, ontology, and the horizon of the Other', *Indo-Pacific Journal of Phenomenology*, 10:2, 23–33.

Porteous, J. D. and Smith, S. E. (2001), *Domicide: The Global Destruction of Home*, Montreal and Kingston: McGill-Queens University Press.

Preston, B. L., Dow, K. and Berkhout, F. (2013), 'The climate adaptation frontier', *Sustainability Science*, 5:3, 1011–35.

Pugliese, J. (2002), 'Penal asylum: refugees, ethics, hospitality', *Borderlands*, 1.1, <http://www.borderlands.net.au/vol1no1_2002/pugliese.html> (last accessed 29 September 2017).

Ramlogan, R. (1996), 'Environmental refugees: a review', *Environmental Conservation*, 23:5, 81–8.

Rasmussen, S. E. (2016), 'First wave of Afghans expelled from EU states under contentious migration deal', *The Guardian*, 16 December, <https://www.theguardian.com/global-development/2016/dec/15/first-wave-afghans-expelled-eu-states-contentious-migration-deal-germany-sweden-norway> (last accessed 22 December 2016).

Reckdahl, K. (2006), 'Desert Storm', *Phoenix NewTimes*, 13 July, <http://www.phoenixnewtimes.com/2006-07-13/news/desert-storm/> (last accessed 23 December 2016).

Reinert, E. S. (2007), *How Rich Countries Got Rich . . . and Why Poor Countries Stay Poor*, London: Constable.

Reuveny, R. and Moore, W. H. (2009), 'Does environmental degradation influence migration? Emigration to developed countries in the late 1980s and 1990s', *Social Science Quarterly*, 90:3, 461–79.

Rio Declaration on Environment and Development 1992, entered into force 22 December, 1992.

Risse, M. (2009), 'The right to relocation: disappearing island nations and common ownership of the earth', *Ethical and International Affairs*, 23:3, 281–300.

Roberts, A. R. (ed.) (2005), *Crisis Intervention Handbook: Assessment, Treatment and Research*, London: Oxford University Press.

Roberts, E. (2012), *Loss and Damage in Vulnerable Countries Initiative: Bangladesh Leading the Way on Loss and Damage*, Bangladesh: International Centre for Climate Change and Development.

Roberts, E. and Pelling, M. (2016), 'Climate change-related loss and damage: translating the global policy agenda for national policy processes', *Climate and Development*, 29, 1–14.

Rose, M. (2012), 'Dwelling as marking and claiming', *Environment and Planning D: Society and Space*, 30:5, 757–71.

Rosello, M. (2001), *Postcolonial Hospitality: The Immigrant as Guest*, Stanford: Stanford University Press.

Royal Geographical Society (n.d.), *The Global North/South Divide*, brochure, <https://www.rgs.org/NR/rdonlyres/6AFE1B7F-9141-472A-95C1-52AA291AA679/0/60sGlobalNorthSouthDivide.pdf> (last accessed 23 February 2017).

Sassen, S. (1999), *Guests and Aliens*, New York: New Press.

Saunders, W. G. and Arthur, W. S. (2001), 'Autonomy rights in Torres Strait: from whom, for whom, for or over what?', Discussion Paper No. 215, Canberra: Centre for Aboriginal Economic Policy Research, Australian National University.

Schroeder, W. R. (2013), 'Continental ethics', in H. LaFollette and I. Persson (eds), *The Blackwell Guide to Ethical Theory*, Malden, MA and Oxford: Wiley-Blackwell, pp. 461–86.

Schlunke, K. (2009), 'Home', *South Atlantic Quarterly*, 108:1, 1–26.

Schuemer-Cross, T. and Heaven Taylor, B. (2009), *The Right to Survive: The Humanitarian Challenge for the Twenty-First Century*, Oxford: Oxfam International.

Scott C. H. and Mulrennan M. E. (2010), 'Reconfiguring *Mare Nullius*: Torres Strait Islanders, indigenous sea rights and the divergence of domestic and international norms', in M. Blaser, R. de Costa, D. McGregor and W. D. Coleman (eds), *Indigenous Peoples and Autonomy: Insights for a Global Age*, Vancouver: UBC Press, pp. 148–76.

Seltzer, M. (1997), 'Wound culture: trauma in the pathological public sphere', *October*, 80, 3–26.

Semuels, A. (2015), 'The village that will be swept away', *The Atlantic*, 30 August, <http://www.theatlantic.com/business/archive/2015/08/alaska-village-climate-change/402604/> (last accessed 5 September 2016).

Serrat, O. (2008), 'The sustainable livelihoods approach', *Knowledge Solutions*, Manila: Asian Development Bank.

Sevoyan, A. and Hugo, G. (2015), 'Vulnerability to climate change among disadvantaged groups: the role of social exclusion', in J. P. Palutikof, S. L. Boulter, J. Barnett and D. Rissik (eds), *Applied Studies in Climate Adaptation*, Chichester: John Wiley & Sons, pp. 258–65.

Shamsuddoha, Md, Khan, Munjurul Hannan Khan, S. M., Raihan S. and Hossain, T. (2012), *Displacement and Migration from Climate Hot-Spots in Bangladesh: Causes and Consequences*, Dhaka: Centre for Participatory Research and Development and ActionAid.

Shaw, L., Edwards, M. and Rimon, A. (2014), *KANI Independent Review Report*, Brisbane: Griffith University.

Shue, H. (1999), 'Global environment and international inequality', *International Affairs*, 531–45.

Siddiqui, T. (2010), 'Mainstreaming migration in national development strategies', background paper for *Symposium on Overcoming Barriers: Building Partnerships for Migration and Human Development*, 27–8 May 2010, United Nations Development Programme, Geneva, Switzerland, <http://ndc.gov.bd/lib_mgmt/webroot/earticle/2111/Labour_Migration_From_Bangladesh_2012.pdf> (last accessed 27 December 2016).

Siddiqui, T. and Sultana, M. (2012), *Labour Migration from Bangladesh 2012: Achievements and Challenges*, Dhaka: Refugee and Migratory Movements Research Unit.

Simonet, G. (2010), 'The concept of adaptation: interdisciplinary scope and involvement in climate change', *Sapiens*, 3:1, <https://sapiens.revues.org/997> (last accessed 27 December 2016).

Skeldon, R. (2008), 'International migration as a tool in development policy: a passing phase?', *Population and Development Review*, 34:1, 1–18.

Soguk, N. (1999), *States and Strangers: Refugees and Displacements of Statecraft*, Minneapolis and London: University of Minnesota Press.

Sokoloff, W. W. (2005), 'Between justice and legality: Derrida on decision', *Political Research Quarterly*, 58:2, 341–52.

Sommers, S. R., Apfelbaum, E. P., Dukes, K. N., Toosi, N. and Wang, E. J. (2006), 'Race and media coverage of Hurricane Katrina: analysis, implications, and future research questions', *Analyses of Social Issues and Public Policy*, 6:1, 39–55.

Stabinsky, D. (2012), *Tackling the Limits to Adaptation: An International Framework to Address 'Loss and Damages' from Climate Change Impacts*, ActionAid International, Care International and WWF International.

Stockholm Declaration on the Human Environment 1972, in *Report of the United Nations Conference on the Human Environment*, UN Doc.A/CONF.48/14, at 2 and Corr.1, <http://www.un-documents.net/aconf48-14r1.pdf> (last accessed 26 October 2017).

Stone, C. D. (2010), *Should Trees Have Standing? Law, Morality and the Environment*, Oxford and New York: Oxford University Press.

Swain, A. (1996), 'Environmental migration and conflict dynamics', *Third World Quarterly*, 17:5, 959–73.

Sydney Morning Herald (2011), 'We're all Queenslanders now', *Sydney Morning Herald*, 15 January, <http://www.smh.com.au/opinion/editorial/were-all-queenslanders-now-20110114-19r5o.html?skin=text-only> (last accessed 23 December 2016).

Tabucanon, G. M. (2012), 'The Banaban resettlement: implications for Pacific environmental migration', *Pacific Studies*, 35:3, 1–28.

Tacoli, C. (2009), 'Crisis or adaptation? Migration and climate change in a context of high mobility', *Environment and Urbanisation*, 21:2, 513–25.

Taran, P. A. (2001), 'Human rights of migrants: challenges of the new decade', *International Migration*, 38:6, 7–51.

The Age (2011), 'Queensland estimates flood bill to top $5b', *The Age*, 28 January, <http://www.theage.com.au/business/queensland-estimates-flood-bill-to-top-5b-20110128-1a7ou.html> (last accessed 23 December 2016).

The Australian (2011), 'Experts predict 50 million "environmental refugees" by 2020', *The Australian*, 22 February, <http://www.theaustralian.com.au/news/world/experts-predict-50-million-environmental-refugees-by-2020/story-e6frg6so-1226009927584> (last accessed 7 December 2016).

Torres Strait Regional Authority (TSRA) (n.d.), 'Community profiles', <http://www.tsra.gov.au/the-torres-strait/community-profiles> (last accessed 9 December 2016).

Torres Strait Regional Authority (TSRA) (2009), *Torres Strait & Northern Peninsula Area Regional Plan: Planning for Our Future: 2009 to 2029*, Thursday Island, Queensland: Torres Strait Regional Authority, <http://www.tsra.gov.au/__data/assets/pdf_file/0018/1773/ts-npa-rp-09-29.pdf> (last accessed 27 December 2016).

Torres Strait Regional Authority (TSRA) (2010), *Torres Strait Climate Change Strategy 2010–2013*, report prepared by the Environmental Management Program, Thursday Island, Queensland: Torres Strait Regional Authority.

Torres Strait Regional Authority (TSRA) (2012), 'Climate change and possible relocation of Island communities in the Torres Strait', 4 October, media release no. 474, <http://www.tsra.gov.au/__data/assets/pdf_file/0009/2889/Media-Release-474-Climate-Change-Possible-Relocation-Torres-Strait-Island-Communities.pdf> (last accessed 27 December 2016).

Torres Strait Regional Authority (TSRA) (2014), 'TRSA response to 2014 federal budget', 15 May, media release no. 559, <http://www.tsra.gov.au/__data/assets/pdf_file/0018/6237/MR-559-TSRA-Response-to-2014-Federal-Budget.pdf> (last accessed 27 December 2016).

Tranwith, C. (2011), 'Tent metropolis for flood refugees', *Sydney Morning Herald*, 18 January, <http://m.smh.com.au/environment/weather/tent-

Bibliography

metropolis-for-flood-refugees-20110118-19uwy.html?page=2> (last accessed 23 December 2016).
Trevillian, J. (2008), 'Taking with ghosts: a meeting with Old Man Crocodile on Cape York Peninsula', *Australian Humanities Review: Ecological Humanities*, 45.
Trombetta, M. J. (2008), 'Environmental security and climate change: analysing the discourse', *Cambridge Review of International Affairs*, 21:4, 585–602.
United Nations Convention Relating to the Status of Refugees 1951, entered into force 28 July 1951, Geneva, Switzerland.
United Nations Declaration on the Rights of Indigenous Peoples 2007, opened for signature 29 June 2006, entered into force 13 December 2007.
United Nations Department of Economic and Social Affairs (2015), *World Population Prospects: The 2015 Revision*, New York: Department of Economic and Social Affairs: Population Division, United Nations.
United Nations Framework Convention on Climate Change (n.d.), 'Adaptation', <http://unfccc.int/focus/adaptation/items/6999.php> (last accessed 21 October 2017).
United Nations Framework Convention on Climate Change 1992, opened for signature June 1992, entered into force 21 March 1994.
United Nations Framework Convention on Climate Change 1997, Kyoto Protocol to the United Nations Framework Convention on Climate Change, adopted at COP3 in Kyoto, Japan, 11 December 1997.
United Nations General Assembly (2016), *In Safety and Dignity: Addressing Large Movements of Refugees and Migrants*, Geneva: United Nations General Assembly.
United Nations Habitat (2011), *On Solid Ground: Addressing Land Tenure Issues Following Natural Disasters*, <http://www.oicrf.org/document.asp?ID=9663> (last accessed 24 November 2016).
United Nations High Commissioner for Refugees (UNHCR) (2015), *Global Trends: Forced Displacement in 2015*, <http://www.unhcr.org/576408cd7.pdf> (last accessed 23 November 2016).
United Nations Office of the High Commissioner for Human Rights (2016), 'Migration and human rights', <http://www.ohchr.org/EN/Issues/Migration/Pages/MigrationAndHumanRightsIndex.aspx> (last accessed 23 November 2016).
United Nations Permanent Forum on Indigenous Peoples (n.d.), *Climate Change and Indigenous Peoples*, Backgrounder, <http://www.un.org/en/events/indigenousday/pdf/Backgrounder_ClimateChange_FINAL.pdf> (last accessed 28 December 2016).
Universal Declaration of Human Rights 1948, opened for signature 10 December 1948, entered into force 16 December 1948, Palais de Chaillot, Paris.

Urry, J. (2007), *Mobilities*, Cambridge: Polity Press.
Uvin, P. (2007), 'From the right to development to the rights-based approach: how human rights entered development', *Development in Practice*, 17:4–5, 597–606.
Vanderheiden, S. (2015), 'Justice and climate finance: differentiating responsibility in the Green Climate Fund', *The International Spectator*, 50:1, 31–45.
Vaughan-Williams, N. (2015), *Europe's Border Crisis: Biopolitical Security and Beyond*, London: Oxford University Press.
Vidal, J. (2013), '"We are fighting for survival," Pacific islands leader warns', *The Guardian*, 1 September, <https://www.theguardian.com/environment/2013/sep/01/pacific-islands-climate-change> (last accessed 4 October 2016).
Vincent, A. (2010), *The Politics of Human Rights*, Oxford: Oxford University Press.
Wade, M. (2016), 'Why Australia is stingy and getting stingier', *Sydney Morning Herald*, 31 May, <http://www.smh.com.au/comment/why-australia-is-one-of-the-worlds-stingiest-countries-20160531-gp866s.html> (last accessed 23 November 2016).
Waldholz, R. (2017), 'Alaskan village, citing climate change, seeks disaster relief in order to relocate, *npr.org*, 10 January, <http://www.npr.org/2017/01/10/509176361/alaskan-village-citing-climate-change-seeks-disaster-relief-in-order-to-relocate> (last accessed 15 June 2017).
Walsham, M. (2010), *Assessing the Evidence: Environment, Climate Change and Migration in Bangladesh*, Geneva: International Organisation for Migration.
Warner, K., Kälin, W., Martin, S., Nassaf, Y., Lee, S., Melde, S., Chapuisat, H. E., Franck, M. and Afifi, T. (eds) (2014), *Integrating Human Mobility Issues within National Adaptation Plans*, UN University – Nansen Initiative Joint Policy Brief #2, Tokyo: United Nations University.
Warner, K. and van der Geest, K. (2013), 'Loss and damage from climate change: local level evidence from nine vulnerable countries', *International Journal of Global Warming*, 5:4, 367–85.
Weil, S. (2002), *The Need for Roots*, London and New York: Routledge.
Westing, A. H. (1992), 'Environmental refugees: a growing category of displaced person', *Environmental Conservation*, 19, 201–7.
White, C. (2011), 'Volunteers rally to help ease flood heartache', *ABC News*, 19 January, <http://www.abc.net.au/news/2011-01-19/volunteers-rally-to-help-ease-flood-heartache/1911226> (last accessed 23 December 2016).
White, G. (2011), *Climate Change and Migration: Security and Borders in a Warming World*, London: Oxford University Press.
White, M. and Schwoch, J. (2006), *Questions of Method in Cultural Studies*, Malden, MA: Wiley.

Whitt, L. A. and Slack, J. D. (1994), 'Communities, environments and cultural studies', *Cultural Studies*, 8.1, 5–31.
Wiley, M. (2013), 'After Sandy: New York's "perfect storm" of inequality in wealth and housing', *The Guardian*, 29 October, <http://www.theguardian.com/commentisfree/2013/oct/28/sandy-new-york-storm-inequality> (last accessed 23 December 2016).
Willox, A. C. (2012), 'Climate change as the work of mourning', *Ethics & the Environment*, 17:2, 137–64.
Wilson, E. K. (2013), 'Be welcome: religion, hospitality and statelessness in international politics', in G. Baker (ed.), *Hospitality and World Politics*, Basingstoke: Palgrave Macmillan, pp. 145–73.
Wolvers, A., Tappe, O., Salverda, T. and Schwarz, T. (2015), *Concepts of the Global South – Voices from Around the World*, Cologne: Global South Studies Centre, University of Cologne, <http://kups.ub.uni-koeln.de/6399/1/voices012015_concepts_of_the_global_south.pdf> (last accessed 15 January 2017).
World Bank (2011), 'Economics of adaptation to climate change', <http://www.worldbank.org/en/news/feature/2011/06/06/economics-adaptation-climate-change> (last accessed 1 September 2016).
Xenos, N. (1993), 'Refugees: the modern political condition', *Alternatives*, 18:4, 419–30.
Zetter, R. (2011), *Protecting Environmentally Displaced People: Developing the Capacity of Legal and Normative Frameworks*, Oxford: Refugee Studies Centre.
Ziarek, E. (2001), *An Ethics of Dissensus: Postmodernity, Feminism and the Politics of Radical Democracy*, Stanford: Stanford University Press.

INDEX

Adaptation
  'adequate adaptation', 36–9
  Bangladesh *see* Bangladesh
  climate–development nexus, 24–9, 39
  common but differentiated responsibility, 2–3, 5, 7, 24, 56, 68, 71–4
  cost–benefit analysis, 42, 54–5
  cultural roots, 33–5
  definition, 4, 19–22
  Derrida *see* Derrida, J.
  dwelling, 29, 33, 42, 45, 52, 105
  ethics of, 1–5, 19, 32, 43, 52, 55–9, 78
  evolution, 4, 19–21, 40
  financial costs of, 3
  funding, 3, 24, 26, 32, 41, 54–5, 68, 70
  Heidegger *see* Heidegger, M.
  human mobility, 4, 17, 21–3, 40, 62, 64
  human rights, 33, 36–9, 62
  in-situ (local), 4, 13, 17–22, 24, 28, 35, 57
  Intergovernmental Panel on Climate Change (definition of), 21
  land, 29, 38
  Levinas *see* Levinas, E.
  limits of, 40, 63
  loss, 46
  migration *see* migration as adaptation
  National Adaptation Plans, 24
  place, 27–30, 41
  political construction of, 18, 21, 40, 119
  political participation, 38–9
  Torres Strait Islands *see* Torres Strait Islands

Advisory Group on Climate Change and Human Mobility, 13, 27, 68–70, 129
Africa, 65–6
anthropogenic climate change, 20, 55, 93, 115, 136, 147
Arendt, H.
  politics (concept of), 38, 84–5, 140
  social textures, 30, 39
  sovereignty, 74, 83–5, 97
  world (concept of), 38
Asia-Pacific region, 63, 110–11
Association for Climate Refugees, 92, 95, 177n
Australia, 14, 24–8, 35, 55, 56, 70–1, 110–17, 119–42, 144–5, 148, 153–7, 158, 172n, 175n, 178n

Bangladesh, 4, 14, 61, 69, 89–116
  Association for Climate Refugees, 92, 95
  Australia–Bangladesh, 110–16
  Bangladesh Climate Change Strategy and Action Plan, 93, 100–2
  civil society, 94
  climate–development nexus, 31–2, 93–4, 116
  common but differentiated responsibility, 71–4
  Dhaka, 89, 90, 93, 99–102, 105, 112,
  disasters (impact of), 31, 92–3, 98, 100, 103
  history of, 101, 103, 104–6
  hospitality, 92, 108–10, 115
  housing, land and property (HLP rights), 91–2, 95–6, 101, 116

Index

informal settlements ('slums'), 39, 94
intergenerational responsibility, 73, 106
internal displacement, 98–101
international migration, 102–5, 110–15
land corruption, 95–9
managed migration, 73–4
migration as adaptation, 102–10
pro-poor migration, 110–16
relocation, 97,101, 103
remittances, 113
rural–urban migration, 100–1
being-in-the-world, 48
Brexit, 62, 73, 175n
Brisbane Floods (in Queensland, Australia), 14, 128, 144, 153–8

Cancun Adaptation Framework, 68–9, 71, 104, 117
climate refugees, 64–7, 79, 83, 94, 100–2, 148, 155–9, 172n
CO_2 emissions, 1
common but differentiated responsibility, 2–7, 24, 59, 63, 68–78, 83, 86, 111, 117, 167, 171n
continental theory, 6–8, 52, 56, 59, 62
cost–benefit analysis *see* neoliberalism
crisis (also *krisis*), 42, 69, 84, 102, 115, 131, 141, 145–52, 156, 159, 160, 161, 163, 164, 174n
cultural theory, 11
 methodology, 11–12

deep history, 66–7
Deleuze, G., and Guattari, F., 30
Derrida, J.
 at-home, 9–10, 43, 57–8, 107
 burial, 137–8, 141
 decision, 106–10, 149–52
 deconstruction, 7–12, 58
 dwelling, 9–11, 43, 51–2, 57–8, 90–2, 137, 164, 167
 finitude, 137
 hauntology, 14, 121, 136–8
 hospitality, 7–12, 14, 23, 52–3, 74, 77–8, 84–6, 92, 107–117, 137, 179n
 interrupted subject, 149–51
 negotiation, 57–8, 77, 106–10

ontopology, 18, 23, 30, 43, 57–8, 164
pharmakon, 93
place/lessness, 27, 51, 57, 74, 92
sovereignty, 8–9, 74, 78, 82, 84–6, 141–2
the future, 43
dispossession, 9, 39, 55, 73, 98–9, 101, 106, 122, 124, 128, 131, 136, 141, 147, 160–1, 166
distributive responsibility, 1, 5, 33, 40, 43, 72, 102, 106, 167
domicide, 144, 152–3, 157, 162
 resistance to, 163–4
dualism, 5, 47, 54, 62–3, 79, 80, 86, 119, 140, 148, 165
dwelling *see* Derrida, J.; Heidegger, M.; Levinas, E.

earth-as-dwelling, 168
Earth Jurisprudence, 168
ethics
 care, 6, 43, 45, 59, 78–9, 86, 127, 131, 149, 168
 imagination, 2–5, 52, 166
 Levinas *see* Levinas, E.
 politics, 3, 8, 10, 109, 172n
 violence, 10, 43, 115, 178n
European Union, 62, 80, 165, 174n
existential crisis, 47–8, 130, 148

Giroux, H., 25, 97–8, 146, 160–1
ghosts, 14, 118–21, 136–9, 141–2, 169

hauntology *see* Derrida, J.
Heidegger, M.
 authenticity, 42, 46–51, 58, 130
 being-with (also *Mitsein*), 48, 56
 care, 45–9
 dwelling, 9, 44–8, 166
 mortality, 43, 45, 47, 50, 149, 151
 plight of dwelling, 45–8, 53, 54, 141, 151, 166
 technology, 44–6, 167, 173n
hospitality
 common but differentiated responsibility, 72
 Derrida *see* Derrida, J.
 Kant, E., 7
 Levinas *see* Levinas, E.
 thinking, 12

human rights
  development agenda, 36, 39
  economic, social and cultural, 34, 37, 53, 75, 81–2, 137
  general, 5, 33, 37, 53, 91, 123–4, 152
  health of the environment, 37–8
  housing, land and property (HLP rights), 13, 37, 39, 90–2, 94–9, 102, 116, 163
  indigenous *see* United Nations Declaration on the Rights of Indigenous Peoples 2007
  labour migration, 113
  mobility, 5, 61–2, 66, 68–9, 70–1, 74–7, 79–80, 92, 102, 111, 113, 162, 167–8
  political participation, 36, 38, 39, 53
  poststructuralism, 6–8, 10–11, 13, 76–82, 91–2, 107–9
  protection agenda, 5, 7, 62, 67, 72, 77–82, 94
  refugees, 69, 75, 81–3
Hurricane Katrina
  displacement, 160
  domicide, 144
  hospitality (refusal of), 159–62
  individual responsibility (discourse of), 160
  neoliberalism, 160–2
  race, 146, 159
  refugees, 158–9

immobility
  Bangladesh *see* Bangladesh
  detention centres, 71
  New Orleans, 160
  trapped populations, 62, 166, 172n
impermanence
  groundlessness, 41, 43, 47
  mortality, 151
intergenerational justice, 2, 5, 14, 33, 40, 52, 111, 119, 121
Intergovernmental Panel on Climate Change (IPCC), 21, 26, 31, 64, 67, 121–2, 169, 172n
internal displacement, 5, 39, 62, 98, 101–2
International Covenant on Economic, Social and Cultural Rights 1966, 37, 81–2
  International Organisation of Migration, 4, 17, 63, 64, 85, 92, 103, 176–7n
Inuit Petition, 13, 42, 44, 52–60
inviolability of the home, 44, 52–7

Kant, E., 7
Kiribati, 27, 34, 38, 61, 111, 114–15, 167
Kiribati Australia Nursing Initiative (KANI), 111, 114–15
Klein, N., 1
Kyoto Protocol 1997, 2, 3, 67, 135

Levinas, E.
  dwelling, 9–11, 42–4, 49–52, 58, 76–8, 90–1
  egoism, 50–2
  ethics, 78–9
  hospitality, 50–2, 58, 78–9
  mortality (of the Other), 49
  Other-oriented rights, 74–82, 91, 94, 163
  suffering, 9, 44, 50, 117, 131, 141, 145–9, 154–5, 160, 173n
  the third (also called 'justice'), 10, 60, 76, 79, 109, 149
loss, 14, 22, 27, 38–9, 44, 46, 50–5, 97, 120, 124, 130–2, 135–6, 141, 145–6, 150, 153, 160–3
Loss and Damages, 70, 93, 175–6n, 177n

Malkki, L., 29, 30, 32
Mansfield, N., 77, 84, 107, 109, 112, 139, 144
media witness, 145, 151, 154, 156, 158–9
migration
  Bangladesh *see* Bangladesh
  causality, 7, 17, 63, 67, 72–4
  common but differentiated responsibility, 71–4
  forced, 69, 80–2, 97, 101, 127, 128–9, 135, 177n
  historical constant, 30, 64, 66–7
  middle-class, 112–15
  security, 7, 30, 32, 65–6, 85
  Torres Strait Islands *see* Torres Strait Islands
  migration as adaptation, 4, 18–9, 23–4, 27, 32–3, 35, 63–70, 102–5, 116, 127–9

Index

migration management, 28–9, 30, 70, 72, 82–3, 161, 177n
migration with dignity,70, 85, 105

Nail, T., 13, 23, 63, 66, 67, 82, 83, 85, 165
National Adaptation Plans (NAPS), 24, 68
national roots, 29–30
neoliberalism, 25–6, 34, 39, 42–3, 46, 52–4, 95, 123, 139, 149, 150, 160–3
new normal, 145, 147–8, 151
Newtok (Alaska), 13, 55–61, 130
normative ethics, 2, 5, 6, 8, 10, 18, 40, 43, 52, 59, 66, 72, 74, 79, 82, 102, 110, 112, 167, 177n

Obama, B., 165–6
ontopology *see* Derrida, J.
Other-oriented rights *see* Levinas, E.

Pacific Islands
  adaptation (local), 26, 34–5
  climate impacts, 26
  detention centres, 28
  dwelling, 35
  mobility, 4, 27, 28, 30, 35
  relocation, 4, 27
  rights, 36–8
  roots (cultural), 34
pro-poor migration, 74, 110–16
protection agendas, 5, 7, 62, 69, 72, 75, 77–83, 94, 176n

race, 28, 64, 126, 144, 146–7, 158–63, 166
refugee (political), 17, 29–30, 64, 155–6
  responsibility-sharing (also called burden-sharing), 65, 72, 175n
relocation
  Alaska *see* Inuit Petition; Newtok (Alaska)
  Bangladesh *see* Bangladesh
  Pacific Islands *see* Pacific Islands
  Torres Strait Islands *see* Torres Strait Islands
remittances, 113, 115
resilience, 14, 24, 27–8, 31, 41, 54, 65, 68, 70, 93, 120–2, 124, 125, 127, 129, 145, 147, 164

securitisation, 64, 66, 71, 148–9, 150
sovereignty
  deconstruction, 8–9, 74, 78, 82–6, 108, 139, 167, 179n
  migration, 28–9, 34, 69, 71, 82–6
  nation-state, 5, 28–9, 34, 40, 69, 71, 74, 77, 82–4, 86, 102, 135, 140–2, 167
  subjectivity, 8–9, 18, 76, 86, 167
  violence, 74, 83, 84, 140–2
spectacle
  climate crisis, 147–8, 151
  dwelling, 144, 160–1
  politics, 144, 152
  social relationship, 146, 151–2, 155
  trauma, 146, 151, 161
  violence, 143
Superstorm Sandy (New York), 147

Torres Strait Islands
  adaptation (local), 122, 128, 156
  autonomy, 125, 139–47
  burial sites, 136–7
  climate impacts, 121–2, 153
  colonialism, 119, 125–7, 131–42, 153
  cost–benefit analysis, 123
  dispossession, 131–2
  governance, 125–7, 140–1, 153
  intergenerational ethics, 120
  land rights, 123, 131, 140
  loss, 123, 136–7
  migration, 119, 127–9
  relocation, 129–36
  Saibai Island, 129, 136
  sea rights, 140
  United Nations Declaration on the Rights of Indigenous Peoples 2007, 119, 123, 135, 139
  World War II, 128, 132–3

uncertainty, 55, 99
United Nations Convention Relating to the Status of Refugees 1951, 69, 75, 81, 175n
United Nations Declaration on the Rights of Indigenous Peoples 2007, 119, 123, 135, 139
United Nations Framework Convention on Climate Change 1992, 2, 3, 20–2, 29, 67, 68–9, 72, 123

United Nations High Commissioner for
    Refugees, 69
United States, 52, 54, 62, 145, 148, 158,
    163, 165
Universal Declaration of Human
    Rights, 37, 75

Vulnerability
    environment, 45, 58, 152
    individuals,154
    mobility, 75, 79–80
    mortality, 47, 58–9
    subjectivity, 149
    universalising, 14, 145, 151–64

wound culture, 151, 160–1

xenophobia, 62, 64h

EU representative:
Easy Access System Europe
Mustamäe tee 50, 10621 Tallinn, Estonia
Gpsr.requests@easproject.com

www.ingramcontent.com/pod-product-compliance
Lightning Source LLC
Chambersburg PA
CBHW051116230426

43667CB00014B/2608